VACUUM TECHNOLOGY

VACUUM
TECHNOLOGY

An introduction

by

L. G. CARPENTER

M.A., B.Sc. (Oxon), F.Inst.P.

American Elsevier Publishing Company, Inc.

New York

1970

© L. G. CARPENTER 1970

American edition published by
American Elsevier Publishing Company, Inc.
52 Vanderbilt Avenue
New York, New York 10017

Made and printed in Great Britain

Library of Congress Catalogue Number: 76–19670–6

To E.T.N.

PREFACE

The aim of this book is twofold: first, to provide for those who are about to use vacuum technology as a tool for the first time an introduction to its principles, before they consult more specialized works and the literature, and second, to present a broad view of the field to those who for professional reasons need a general knowledge of the subject.

This double aim has imposed omission of detail, brevity of style and, in some cases, over-generalization, the evils of which I hope may be mitigated by the references given to more specialized books and papers.

Appendices I and II are intended for the convenience of those readers who may wish to be reminded of the physical formulae and terms used in the text.

I gratefully acknowledge my indebtedness to many friends and colleagues, including Mr G. I. Lewis, Mr W. N. Mair, Mr B. D. Power, Mr T. W. G. Rowe, Mr P. Ridgway Watt, and particularly Dr G. W. Green, who read the whole of the first draft and made many constructive criticisms and suggestions. Mr M. J. Watts kindly read the proofs, but responsibility for remaining errors is entirely mine.

My thanks are due to Professor J. P. Roberts, whose invitation to lecture on vacuum technology in the Department of Ceramics at the University of Leeds led to this book, and to Professor D. W. Holder for stimulating discussions and for the hospitality of his laboratory.

To the publishers, and especially to Mr Neville Goodman, whose helpful patience has been unfailing, I am deeply grateful.

L. G. CARPENTER

Department of Engineering Science
Oxford

June 1969

1

VACUA
AND
VACUUM TECHNOLOGY

§ 1.1 What is a vacuum?

The term *vacuum* is generally used to denote a portion of space in
which the pressure is significantly less than 760 torr.* The definition is refined by the British Standards Institution as follows:

Medium high vacuum	1 to 10^{-3} torr
High vacuum	10^{-3} to 10^{-7} torr
Very high vacuum	10^{-7} torr and below

From the physical point of view, a high vacuum *situation* exists
where the mean free path (see Appendix I, § 9) of the molecules, *as
calculated from the pressure*, is much greater than the linear dimensions of the vessel. The situation is then dominated by gas/wall (as
opposed to gas/gas) collisions. Viscosity, as ordinarily defined, has
little reference or meaning in these circumstances, the basic concept being one of random wandering from one gas/wall collision to
another without 'memory' of incident direction before collision.
However, one aspect of vacuum technology which does not fall
within this definition of vacuum situations is the action of vapour-
stream (diffusion) pumps whose action depends on gas/gas
collisions. This is dealt with in § 3.3.

* 760 torr = 1 standard atmosphere, and 1 torr is very approximately the
pressure due to a column of mercury 1 millimetre in height. The unit
millitorr (m torr) or micron (μ) is also used.

§ 1.2. The basic topics of vacuum technology

The creation of a vacuum involves two processes, the first being the removal of the gas originally contained in the volume of the work chamber, and the second a competition between the capacity of the pump and the production of gas, not originally in the gas phase, but stemming from leaks—real or virtual, the latter being gas desorbed (see Appendix II, § 4) from the walls and contents.

Hence the topics to be treated include the speed of pumps and the conductance of channels by which they are connected to the work chamber (Chap. 2); their construction and mode of operation (Chap. 3); means of determining the pressures achieved and the composition of the gases composing them (Chap. 4); the design and choice of materials to minimize leaks, both real and virtual (Chap. 5); and the rather special techniques needed for producing and measuring the highest vacua (Chap. 6). Some practical applications of vacuum technology are treated in Chap. 7.

2

CONDUCTANCE
AND
PUMP SPEED

§ 2.1. The concept of conductance

Conductance (F) is concerned with the speed of gas flow through an aperture or channel under unit pressure difference. It is somewhat analogous to electrical conductance in that, as will become evident in § 2.2, the reciprocals of conductances in series (i.e. resistances) are additive.

The usual definition is as follows:

$$F = \frac{\text{flux of gas molecules in } PV \text{ units* per second}}{\text{pressure difference across the aperture or channel}}$$

Thus, if $\dot{V}_1$ is† the volume $\sec^{-1}$ (measured at the pressure P_1) traversing the aperture or channel, across which a pressure difference $P_1 - P_2$ exists,

$$F = \frac{P_1 \dot{V}_1}{P_1 - P_2} \qquad (2.1)$$

For a perfect gas (see Appendix I, equation 8)

$$P_1 \dot{V}_1 = n \dot{V}_1 kT$$

and
$$P_1 - P_2 = (n_1 - n_2)kT \qquad (2.2)$$

---

* For meaning of PV units, see Appendix I, § 7.
† The notation $\dot{x}$ is used throughout for dx/dt.

where n denotes molecules cm^{-3}, k is Boltzmann's constant, and T is the absolute temperature.

Then
$$F = \frac{n_1 \dot{V}_1}{n_1 - n_2} \tag{2.3}$$

$$= \frac{\text{flux of molecules per second}}{\text{difference in molecular density}} \tag{2.4}$$

and has dimensions [Length]^3 [Time]$^{-1}$, i.e. volume per unit time.

§ 2.2. Conductances in series

It is sometimes convenient to work with the impedance (W) which is defined as the reciprocal of F, i.e., if the flux is $\dot{Q}$ (in PV units per unit time)

$$\dot{Q} = \Delta P . F$$

$$= \frac{\Delta P}{W} \tag{2.5}$$

where ΔP is the pressure drop across any one conductance. Consider a number of conductances in series across which the total pressure drop is P. In the steady state $\dot{Q}$ must be the same for each.

For the ith element

$$\Delta P_i = \dot{Q} W_i$$

$$\therefore \sum \Delta P_i = \dot{Q} \sum W_i$$

or
$$P = \dot{Q} \sum W_i$$

But, by definition of W, for the whole assembly

$$P = \dot{Q} W$$

Hence
$$W = \sum W_i \tag{2.6}$$

or
$$\frac{1}{F} = \sum \frac{1}{F_i}$$

The practical consequence of (2.6) is that the overall conductance of a number of conductances in series is dominated by the smallest conductance (largest impedance) of the assembly, in analogy with the dominance of the largest resistance in a chain of series resistances in the electrical case.

Clearly the effective conductance of a number of *parallel* conductances is their sum.

§ 2.3. The effect of pressure on flow regime and conductance

Three regimes of gas flow may be distinguished. At high pressures —λ (the mean free path) much less than d (a typical dimension of the vessel)—there is Poiseuille flow, which is dominated by viscosity, i.e. by inter-molecular collisions. At the other extreme ($\lambda \gg d$) there is free molecular flow, in which gas molecule/wall collisions are dominant. The intermediate case is called transition flow, where both mechanisms are operative.

For molecular flow F is independent of pressure, and dependent only on geometry and on the molecular weight and temperature of the gas; it is therefore a simple and most useful concept. Molecular flow alone will be dealt with here, because it is a truly vacuum phenomenon (in the sense of § 1.1), and because it is the conductance in the low-pressure molecular flow channels which, except in very large chambers or in pumping large continuous gas loads, usually determines the efficiency of the pumping system as a whole.

Conductance in the Poiseuille and transition regimes is a function of pressure as well as geometry and temperature. Values of conductance in the three regimes are given for example by Dennis and Heppell (ref. 2.1) in a form useful for design.

§ 2.4. The conductance of an aperture

Consider the case of molecular flow through an aperture of area A in a diaphragm of area large compared with A, separating two vessels at the same temperature in which the molecular densities are n_1 and n_2 molecules cm^{-3} respectively, and the mean speeds (being determined by the temperature) have a common value $\bar{c}$ cm sec^{-1}. From equation (6) of Appendix I, the numbers of molecules which 'hit' the aperture from the two sides* and pass through it are $An_1\bar{c}/4$ and $An_2\bar{c}/4$ respectively, hence by equation (2.4)

$$F = \frac{A}{4}\bar{c}\frac{n_1 - n_2}{n_1 - n_2}$$

$$= \frac{A\bar{c}}{4} \qquad (2.7)$$

Note that F is directly proportional to area and independent of shape. The validity of the formula depends upon the edges of the aperture being infinitely thin, otherwise some of those molecules which hit the edge of the aperture fail to pass through it, the magnitude of this failure depending on the ratio of the thickness of the aperture walls to its linear dimensions.

At 25°C, $\bar{c}$ for N_2 is $4\cdot745 \times 10^4$ cm sec^{-1}, hence F for an aperture of 1 cm^2 is about $11\cdot9$ litres sec^{-1}. Equation (3) of Appendix I shows that $\bar{c}$ and therefore F is proportional to $(T/m)^{1/2}$, where m is the mass of the molecule.

§ 2.5. The conductance of a channel

For molecular flow through a long channel, ref. 2.2 gives the conductance as

$$F = \frac{4}{3}\bar{c}\frac{1}{\int_0^l (H/A^2)dx} \qquad (2.8)$$

where H is the perimeter, A the cross-section area at any point x, and l is the total length.

*Assuming the Maxwellian speed distribution undisturbed by the flow.

For the particular (and common) case of a long tube of constant radius a, formula (2.8) becomes, if a and l are in cm,

$$F = \frac{2}{3}\frac{\pi a^3 \bar{c}}{l}$$

$$= 30\cdot48\,\frac{a^3}{l}\left(\frac{T}{M}\right)^{1/2} \text{litres sec}^{-1} \tag{2.9}$$

where T is the absolute temperature in degrees Kelvin and M is the molecular weight. A useful approximation for air at 25°C is

$$F = 10^2\,\frac{a^3}{l}\,\text{litres sec}^{-1} \tag{2.10}$$

Equation (2.9) gives the conductance of a long tube, and ignores the fact that, even if the tube were infinitely short (i.e. merely an aperture), it would have (see equation 2.7) a conductance of $\pi a^2 \bar{c}/4$. Remembering that the reciprocals of conductances are additive it may be shown that the total conductance is approximately obtained by adding a length $8a/3$ to the geometric length l of the tube. Thus for short tubes (a comparable with l), equation (2.9) overestimates the conductance. The end effect is shown by the curvature in the large diameter, small length region of Fig. 2.1, which gives curves for the conductance of circular tubes for air at 20°C.

The mechanism of conductance is discussed from the molecular point of view in ref. 2.3, and Steckelmacher (ref. 2.4) has given an excellent critical review of work on molecular flow conductance for systems of tubes and components.

§ 2.6. The conductance of a cold trap

It is often desirable to interpose between a diffusion pump (see § 3.3) and the work chamber being pumped a tortuous *cooled* channel. Its purpose is to condense the working fluid of the pump at a temperature such that its saturated vapour pressure at this temperature is acceptably low in the work chamber. Such a channel is called a 'cold trap'.

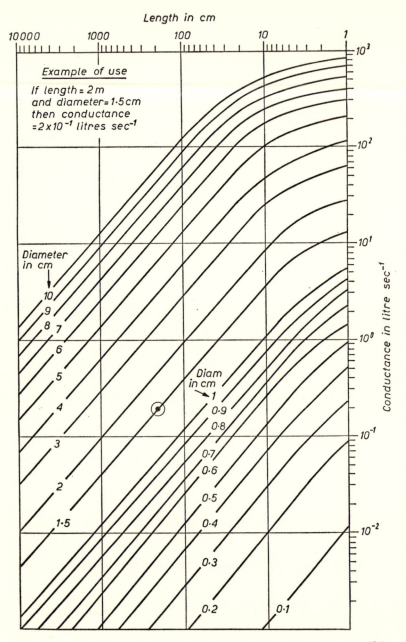

Length in cm

10 000   1000   100   10   1

Diameter in cm
10
9
8 7
6
5
4
3
2
1·5

Example of use

If length = 2 m
and diameter = 1·5 cm
then conductance
= 2 × 10⁻¹ litres sec⁻¹

10^3

10^2

10^1

10^0

10^{-1}

10^{-2}

Conductance in litre sec⁻¹

Diam in cm
1
0·9
0·8
0·7
0·6
0·5
0·4
0·3
0·2   0·1

Fig. 2.1. Conductance of cylindrical tubes for air at 20°C

To be effective, a cold trap must be held at a sufficiently low temperature and must be of such geometry that a vapour molecule of pump fluid cannot get from the pump to the work chamber without hitting the wall of the trap (i.e. the trap is 'optically opaque') with an appreciable probability of sticking to it. Equally, a molecule *of gas to be pumped* cannot traverse the trap from work chamber to pump without also hitting the trap wall and being cooled thereby, i.e. its velocity will be reduced. Thus both opacity and reduction of velocity impede the pumping of gas molecules; so in calculating the total conductance of the pumping system the conductance of the cold trap must be taken into account. The conductance can be computed in some simple cases; see e.g. Dushman and Lafferty (ref. 2.5). In applying formulae valid for air at room temperature, allowance must be made for the effect on $\bar{c}$ of the trap temperature and also of the molecular weight of the gas, if it is other than air.

§ 2.7. The speed of a pump

A pump is a device with an inlet port having the property that a certain fraction of the molecules of the gas to be pumped which enter it do not return. Its speed (S) may be defined as follows:

$$S = \frac{\text{net flow per second entering the pump port, measured in } PV \text{ units}}{\text{pressure } (P) \text{ at the entry port}}$$

$$= \frac{P \times \text{volume sec}^{-1}, \text{ measured at pressure } P}{P}$$

$$= \text{net volume sec}^{-1} \text{ (measured at the entry pressure)}$$
$$\text{traversing the entry port} \quad (2.11)$$

The precise definition of P is discussed in § 3.34. If A is the area of the entry port, and v is the average velocity of traversing the entry port,

$$Av = S \qquad (2.12)$$

or
$$v = \frac{S}{A} \qquad (2.13)$$

= pump speed per unit area of
entry port

If every molecule which crossed the plane of the entry port and entered the pump failed to return, the maximum possible value of v would be $\bar{c}/4$ [equation (6) Appendix I] so that the speed of a pump for air at 25°C could not exceed $\frac{1}{4} \times 4 \cdot 75 \times 10^4$ cm^3 per second per cm^2 area of entry port, i.e. $11 \cdot 9$ litres sec^{-1} cm^{-2} of entry port. If, as is frequently the case, there is a channel of conductance F between the entry port of the pump and the work chamber, and if the flow is in the molecular regime, so that the flux through F is linear in the pressure differential across it, then

$$PS = (P_c - P)F = \text{flux in } PV \text{ units}$$

$$= S_c P_c$$

where P = pressure at entry port of pump

S = speed of pump

P_c = pressure in work chamber

S_c = effective pump speed *at the work chamber*

Algebraic elimination of the pressures from the above equations leads to

$$S_c = S \frac{F}{F+S} \qquad (2.14)$$

Hence, the effective speed S_c is always less than S, the speed of the pump itself, and this reduction can be serious; e.g. if $F = S$, the introduction of F between the pump and the work chamber will halve the effective pumping speed.*

---

* S, the speed of a pump, is a quantity of the same nature as F, the conductance of a channel. In the former, $S \times P$ [which may be written as $S(P - o)$], is the net flux into the pump, in the latter $F(P_1 - P_2)$ is the net flux through the channel.

3

VACUUM PUMPS

§ 3.1. Preliminary survey of types of vacuum pump

According to their mode of operation, pumps may be classified as:

(a) *Mechanical pumps* in which gas is trapped, compressed and removed bodily from the low-pressure to the high-pressure side of the pump, whence it is expelled to the atmosphere either directly or through a second mechanical 'backing' pump.

(b) *Mechanical pumps* which impart velocity to the pumped gas molecules by impact with swiftly moving solid surfaces. Such pumps are always backed by another mechanical pump interposed between them and the atmosphere.

(c) *Vapour stream pumps* in which the gas molecules to be pumped are caused to move in the desired direction by impact with heavy molecules of pump fluid, which derive their velocity by boiling from a liquid reservoir. With the exception of steam ejector pumps (which exhaust direct to atmosphere) vapour stream pumps are backed by mechanical pumps.

(d) *Chemical pumps* in which the molecules to be removed from the work chamber are caused to combine chemically with highly reactive substances ('getters'), such as titanium, thus being converted to and trapped in the solid phase. The getters are dispersed on the pump surfaces by evaporation or by sputtering.

(e) *Sorption pumps* in which the molecules to be pumped are physically sorbed (e.g. by Van der Waals forces) on specially prepared substances of large specific area (e.g. charcoal or zeolites). The sorption is usually aided by cooling the sorbent below room temperature and is reversible with temperature.

(*f*) *Cryo pumps*, which are vessels whose internal chemically non-reactive metal surfaces are cooled by refrigerants to a temperature such that the gases to be pumped condense upon them. After the condensation of one or two monolayers, the process is indistinguishable from the condensation of a substance on a substrate *of itself*, thus differentiating it from (*e*) above.

In pump types (*a*), (*b*), and (*c*), the pumped gas is eliminated from the system; in the other types it is retained.

§ 3.2. Mechanical pumps

3.2.1. *Mechanical oil-sealed pumps*

These pumps, which are very widely used, belong to category (*a*) above. They run immersed in oil, which lubricates and seals the rubbing interfaces and reduces the dead spaces at the end of the compression, thus improving the compression ratio.

The rotating vane type, common for smaller pumps is illustrated in Fig. 3.1. Gas entering the pump from the work chamber is trapped and compressed by the rotation of rotor 3 in stator 4. Gas tightness of the rotor/stator interfaces is maintained by the vanes 1 and 2 (forced outwards by the spring 5 and by centrifugal force) and by the small clearance between the two ends of the rotor and the stator, the seal being improved by the small quantity of oil on the stator and rotor surfaces.

The trapped gas is conveyed to the compression side of the pump and is expelled, together with a little oil, when the pressure exceeds atmospheric sufficiently to raise the exit flap valve 6.

If the gas to be pumped contains a condensable vapour, contamination of the oil and reduction of attainable vacuum result. As a specific example, consider the pumping of air which contains some water vapour. As the inlet pressure to the pump decreases during pump-down, the compression ratio necessary to raise the pressure

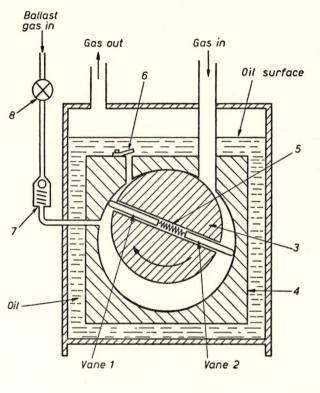

Fig. 3.1. Oil-sealed rotary vane pump, gas ballasted

1	Vane	5	Spring
2	Vane	6	Flap valve
3	Rotor	7	Spring-loaded valve
4	Stator	8	Valve

of the trapped gas to a value sufficient to open the exit valve against the combined resistance of its spring and the atmosphere increases. If the compression ratio is such that the partial pressure of the water vapour is raised to a value equal to the saturated vapour pressure of water at the temperature of the pump, the vapour will condense and may not be completely expelled or will be expelled, via the exit valve, *as liquid* into the oil. Thence it will find its way back into the body of the pump (via the oilways provided for maintaining the lubrication and oil seal) and will re-evaporate, thus setting a limit to the vapour pressure of water in the pumped chamber.

The device of *gas ballast* reduces this unwanted effect by introducing atmospheric air at a point in the cycle (such as that shown by the rotor position in Fig. 3.1) where the gas being pumped is isolated from the inlet port, but is connected to the spring-loaded valve 7. This valve admits air the compression of which, by further rotor movement, is sufficient to lift the exit valve 6 and eject ballast air, gas to be pumped, and water vapour while the latter is still in the vapour phase and can pass right out of the pump. The amount of air ballast is adjustable by valve 8. An excellent account of the theory and practice of gas ballasting is given by Power (ref. 3.1), whose book should be consulted for an authoritative account of oil-sealed rotary pumps and, indeed, pumps of all types.

A typical single-stage rotary pump pumping air might have an ultimate pressure of 5×10^{-3} torr, as measured by a gauge sensitive only to the permanent (non-condensable) gases, but nearly an order of magnitude higher, if the gauge is sensitive to the vapour pressure of the oil and its decomposition products. Its speed would start significantly to decrease at about 1 torr. For this reason, two rotary pumps (constituting a *two-stage* pump) are often put in series and mounted as one unit. Typically, the speed of such a unit might be 1 litre sec^{-1} when driven at 650 rpm by a $\frac{1}{3}$ h.p. motor, decreasing significantly below about 10^{-2} torr, and becoming zero at its ultimate pressure (for permanent gases) of 10^{-4} torr. With gas ballast, speed decrease sets in at a higher pressure, and the ultimate pressure is increased by (typically) a decade.

3.2.2. *Roots pumps*

These pumps, also called Mechanical Booster Pumps, belong, as is evident from Fig. 3.2, to category (*a*) of § 3.1, since gas is bodily removed by the contra-rotating figure-of-eight-shaped rotors from

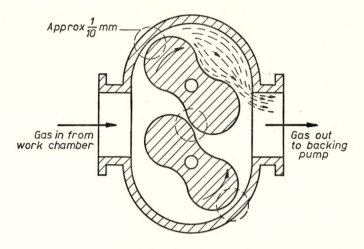

Approx $\frac{1}{10}$ mm

Gas in from
work chamber

Gas out
to backing
pump

Fig. 3.2.   Mechanical booster [Roots] pump
(Courtesy of Leybold Heraeus Ltd)

the low-pressure to the high-pressure side. A mechanical booster
pump is always backed by another pump interposed between it and
the atmosphere. Hence the pressure difference across it is much
less than 760 torr, and oil (with its usually undesirable vapour
pressure) is not necessary for sealing the $\sim 10^{-1}$ mm clearance between rotors and stator, but only for lubricating the gears, which
ensure the necessary very accurate synchronization of the rotors.
These pumps are suitable for high-speed pumping in the pressure
range 10^{-2} to 10 torr. Their characteristics and use are fully discussed by Power (ref. 3.2) and speeds range from about 50 litres
sec^{-1} to more than 100 times this value.

3.2.3. *Mechanical molecular pumps*

These pumps fall into category (*b*) of § 3.2, and depend on the
impingement of molecules on a moving surface and their re-
emission having acquired, superimposed on their Maxwellian
velocities, some of the macroscopic velocity of the moving surface.
Suitable ducting arrangements direct the flow of molecules from

the low-pressure to the high-pressure side of the pump. Two examples are given below.

3.2.3.1. *Molecular drag pumps*

These were introduced in 1913 by W. Gaede, the principle of whose pump is shown schematically in Fig. 3.3. Gas entering port

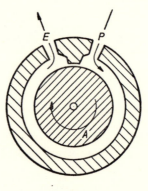

A   Rotor
P and E   Ports

Fig. 3.3. Principle of Gaede's molecular pump

P is expelled at port E, having acquired a mean velocity in a clockwise direction by repeated impact with the surface of rotor A, moving at about 5000 cm sec^{-1}. The subsequent development of rotary drag pumps by Holweck, Siegbahn, and Gondet is described by Power (ref. 3.3) who points out that their most suitable applications exploit either their immediate availability (without warm-up period) and comparatively vapour-free vacuum or their characteristic of pumping heavy molecules better than light ones.*

* Compare diffusion pumps whose speed for lighter gases often exceeds that for heavier ones, because the smaller thermal velocities, and hence diffusion, of heavy molecules causes a lower rate of entrainment (see § 3.3.1.).

A disadvantage of these pumps is the necessarily small clearance between rotor and stator. The turbo-molecular pumps described below have clearances of several millimetres.

3.2.3.2. *Turbo-molecular pumps*

A typical pump (ref. 3.4) consists (Fig. 3.4) of a metal housing in which runs a rotor at 16000 rpm, supported on water-cooled oil-

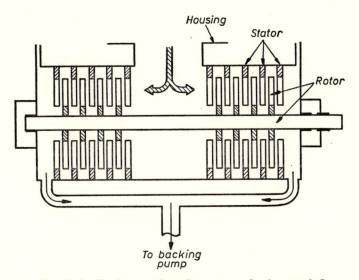

Fig. 3.4. Turbo-molecular pump [schematic]
(From *Vacuum*, 1966. Courtesy of Pergamon Press Ltd)

lubricated bearings at each end of the shaft. Obliquely slotted discs are arranged alternately on the rotor and stator. These are analogous to the rotor and stator blades of a turbine but are simpler in that, at low pressures, aerodynamic shape is not significant. The rotation of the rotor blades urges the gas from the central entry port to each end of the pump whence it flows to the exit port, carrying with it any oil molecules from the lubricated bearings which would otherwise backstream to the work chamber.

Each pair of discs forms a pressure stage, so that, with a backing pressure of 10^{-2} torr, a partial pressure of air of order 10^{-10} torr is achieved. With a speed of 140 litres sec^{-1} the power consumption is about $\frac{1}{3}$ kilowatt. The pressure ratio increases with the molecular weight of the gas pumped (hence the control of gear oil backstreaming mentioned above). The fact that heavier gases are differentially pumped may be useful in some cases (e.g. in thermonuclear research), but in others (e.g. analysis of residual gases by mass spectrometer) a source of difficulty.

Turbo-molecular pumps are mainly used at present for research purposes where organic-free ultra-high vacuum is required without the complication and reduction of pumping speed, associated with L.N$_2$ cooled traps.* It is early to assess their industrial importance. In ref. 3.4, pumps are described with speeds for air ranging from about 140 to 5200 litres sec^{-1}.

§ 3.3. Vapour-stream pumps [Diffusion pumps]

3.3.1. *Basic principle*

A liquid of high molecular weight boils under reduced pressure (provided by the 'backing pump') and by suitable geometry is formed into a vapour jet into which the gas molecules to be pumped diffuse, to be carried by the vapour jet towards the backing pump, whence they are expelled to the atmosphere. Pumps based on this principle are commonly called 'diffusion' pumps, since the molecules to be pumped diffuse into the jet of pump fluid vapour. If a stream of vapour molecules expands through a nozzle, it attains supersonic velocity if the pressure in the region into which it expands is sufficiently low. A number of nozzles are arranged in series so that each jet produces a reduction in pressure and provides the low pressure into which the succeeding jet expands.

The construction is shown schematically in Fig. 3.5, in which the pump fluid, boiling under reduced pressure from the backing pump, passes up the cylindrical chimney, some emerging from the

* L.N$_2$ is a common abbreviation for 'liquid nitrogen'.

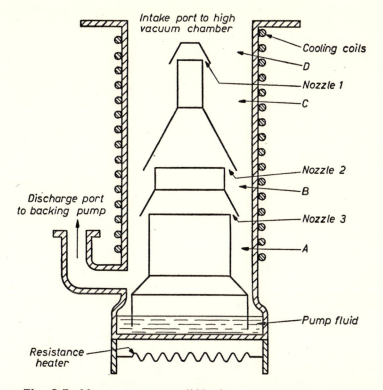

Fig. 3.5. Vapour-stream diffusion pump [schematic]

annular nozzle 3 and carrying with it gas which has diffused from the stage 2 above it, thereby making the pressure at B less than that at A. The actions of nozzles 2 and 1 are similar, the reduction of pressure by a nozzle being typically several orders of magnitude.

The pump fluid must be removed and returned to the boiler after it has performed its function, hence the water cooling of the pump casing. As each stage must handle the same number of molecules and the pressure decreases from A to D, the annular space between nozzle and pump casing increases in the same direction.

Diffusion pumps are often fitted, immediately above the inlet

3

port, with a baffle which is intended to ensure that molecules of pump fluid migrating from the pump to the work chamber (i.e. going in the 'wrong' direction) strike the baffle, are condensed upon it and returned as liquid to the pump. Thus the entry of pump fluid into the work chamber is minimized.

Baffles, like cold traps (see § 2.6) are a compromise between opacity to pump fluid going to the work chamber and opacity to gas molecules coming from the work chamber—the latter being an undesired penalty. They are usually water-cooled, or refrigerated to about $-30°C$, and about halve the speed of the pump to which they are fitted. Baffles very deeply refrigerated with the intention of eliminating as completely as possible the passage of fluid from pump to work chamber are in effect cold traps.

3.3.2. *Mercury diffusion pumps*

The outstanding advantage of this (the earliest type of diffusion pump) is that the working fluid being an element cannot decompose in the boiler (in which the temperature may be about $150°C$); a major disadvantage is the high vapour pressure at the temperature of the cooling water ($\sim 10^{-3}$ torr at $15°C$), so that, unless a cold trap is used, the work chamber is flooded with mercury vapour. But using a trap impedes the passage of gas between the work chamber and the top nozzle and some loss of pump speed results. Mercury-filled pumps are however particularly suitable for systems in which *organic* contamination must be avoided and the use of sputter ion pumps (see § 3.4.1) is not desired. For details see ref. 3.5. Backing is sometimes by water-jet 'filter' pumps.

3.3.3. *Oil diffusion pumps*

The working fluid is an organic liquid (often of molecular weight of order 500 mass units) having a vapour pressure, at the cooling water temperature, in the range 10^{-6} to 10^{-10} torr, depending on species, and boiling under a backing pressure of order 10^{-1} torr in temperature range 150 to $230°C$. For many applications, the presence of low residual oil pressures is acceptable and the use of an additional refrigerated vapour trap, with its attendant in-convenience and decrease of effective pumping speed, is un-necessary. If however a properly designed L.N$_2$-cooled vapour trap is introduced between pump and work chamber, pressures of 10^{-10} torr or less can be attained.

The disadvantage of organic pump fluids is that they tend in service to decompose into components whose room-temperature vapour pressure is much higher than that of the parent pump fluid. They are particularly prone to decomposition if exposed to a high pressure of air, caused for instance by the accidental stoppage of the backing pump or a sudden large leak.

The term *oil* to denote the working fluid is a loose one and covers a large range of organic fluids of which the two most commonly used in Great Britain are probably paraffinic hydrocarbons or silicones, the latter being more stable than the former if suddenly exposed to a high air pressure when hot.

The choice between mercury and the various organic fluids obviously depends on the particular use contemplated, and on the pros and cons, considered in the light of an authoritative account such as given in Power's book. However, a rough generalization is that an 'oil' diffusion pump is usually to be preferred unless there are special considerations favouring mercury. Oil 'booster' pumps (described in Power's book) are similar *in principle* to diffusion pumps.

3·3·4. *The speed of diffusion pumps*

In § 2.7 the speed S of pump is given as the ratio of the net throughput $\dot{Q}$ to the pressure P at the entry port, i.e.

$$S = \frac{\dot{Q}}{P} \qquad (3.1)$$

However, P must be unambiguously defined, there being two conventions and the difference between them becoming significant at low pressures. The pressure at the entry port of the pump is the sum of the partial pressure due to the pump fluid P_u, (called its *ultimate pressure*) and the partial pressure P_g of the gas being pumped.

That is the total pressure P_0 at the entry port is

$$P_0 = P_g + P_u \qquad (3.2)$$

The value of speed obtained by putting $P = P_g$ in (3.1) may be called the *test gas speed* S and there is no reason to suppose that in a well-designed multi-stage diffusion pump $\dot{Q}$ should not remain proportional to P_g at very low pressures, so that the test gas speed, denoted by S, remains constant. Alternatively, if P_0 is inserted in the denominator of (3.1) the *practically measured speed* S' is obtained. Since the test gas flow $\dot{Q}$ is independent of how P is defined, we have

$$\dot{Q} = S'P_0 = SP_g \qquad (3.3)$$

which, inserting (3.2), becomes

$$S' = S\left(1 - \frac{P_u}{P_u + P_g}\right) \qquad (3.4)$$

Hence as the partial pressure (P_g) of the gas being pumped tends to zero, S' tends to zero, but S remains constant.

At the high pressure end of the inlet pressure range, the speed (there being here no significant difference between S and S' since $P_g \gg P_u$) also decreases, probably because the critical backing pressure of one of the stages of the pump is exceeded. Fig. 3.6 illustrates both the above discussed phenomena.

The throughput of a pump, being its rate of removing gas *measured in PV units* per unit time is given at each inlet pressure by the product (pressure × speed). This product is a maximum at the high-pressure end of the speed/pressure curve, usually in the 10^{-1} to 1 torr region. Examination of Fig. 3.6 shows, for example, that the throughput decreases from 10 torr litres sec^{-1} at 10^{-1} torr to $1 \cdot 4 \times 10^{-4}$ torr litres sec^{-1} at 10^{-7} torr.

In § 2.7 it is stated that the maximum speed of a pump for air at 25°C cannot exceed $11 \cdot 9$ litres sec^{-1} per cm^2 of entry port. In most diffusion pumps only about $\frac{1}{3}$ or $\frac{1}{4}$ of the molecules which cross the mouth entry plane are entrained in the vapour stream never to return, the proportion so entrained being called the

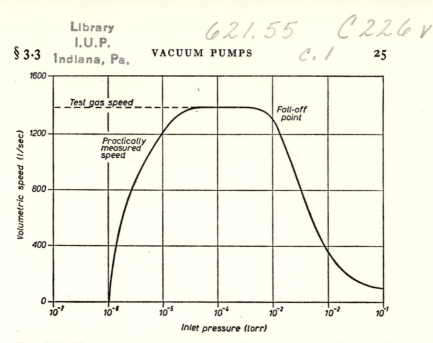

**Fig. 3.6. Test-gas speed and practically measured speed
curves for a diffusion pump of area about 410 cm²**

(From B. D. Power, *High Vacuum Pumping Equipment.* Courtesy of Chapman & Hall Ltd)

'Speed Factor', which for example in Fig. 3.6 has the value
$(1350/410 \times 1/11 \cdot 9)$ of about 0·28 at 10^{-4} torr.

3·3·5. *Backstreaming*

Backstreaming is a measure of the tendency of a pump to contaminate with its pump fluid the work chamber it is pumping.

To fix ideas consider an oil diffusion pump with condensing
walls c at temperature T_c, the mouth being blanked off by a plate
P, held at a constant temperature T_p (equal to T_c) by external
means, i.e. *not* by condensation of pump fluid upon it.

If the whole of the pump between c and P were isothermal, and
if there were no transfer of oil, except by evaporation from oil-
covered surfaces, once P had become covered with an oil film
there would be no further net oil transfer between the pump and P.
Such a pump might be considered an 'ideal' pump, and if P were
removed and the pump connected to a work chamber w, also at T_c,

there would be no continuous oil transfer after an initial wetting of the walls of w.

However, *ideal* pumps do not exist; in *real* pumps there are oil-covered regions warmer than T_c and regions from which oil vapours emanate at rates greater than would be appropriate to evaporation from liquid-covered surfaces at T_c. These regions are mainly associated with the top-stage nozzle of the pump. For simplicity, imagine that there is just one such region L, oil covered (e.g. the lip of the top-stage nozzle) with temperature T_L greater than T_c and hence than T_p, imagined to be maintained equal to T_c.

There will now be a net distillation flux of oil from L to P and its measured value in mass per unit area of pump mouth is called the backstreaming rate B, though the backstreaming flux is not uniform over the area of the pump mouth.

If the blanking off plate P (used in measuring B) is removed from such a real pump and the pump is connected to w (maintained at T_c as before) there would now be a continuous net flow from L to w, and the value of B, multiplied by the mouth area of the pump would be a measure of its value. Thus B is a measure of the magnitude of the net flux of oil into a chamber at temperature T_c from a real pump with condensing walls at T_c, since that from an ideal pump, in which there were no sources from which oil could reach w except oil-covered surfaces at T_c, would be zero.

The backstreaming of a pump is combated if necessary by baffles (§ 3.3.1) and cold traps (§ 2.6). Backstreaming rates depend both on the design of the pump and the vapour pressure of the oil. Typical values of B are from 0·01 to 0·05 mg min^{-1} cm^{-2} for well-designed pumps. This figure can be grossly exceeded in a poor design, while a *guard ring* or *cold cap* over the top-stage nozzle may make it 10–20 times smaller. The above treatment is somewhat idealized and intended merely to expose some of the basic elements of the phenomenon. For full details and methods of measurement see Power (ref. 3.17) to whom I am indebted for illuminating discussions.

The user of diffusion pumped chambers may be interested in one or both of two quantities, viz (*a*) the concentration of oil vapour in the *volume* of the chamber and (*b*) the rate of continuous build up of oil on *surfaces* within the chamber. Effect (*a*) is in principle unavoidable, but can be reduced to any desired level by an appropriately cooled trap (see §§ 2.6 and 3.3.2) between the pump and the chamber. Effect (*b*) can be prevented without the use of traps, by heating the surface to be protected, provided that the necessary temperature is not so high as to decompose the oil into less volatile constituents which 'sit' on the surface.

3.3.6. *Steam ejector pumps*

In these a high-speed jet of steam entrains the gas to be pumped and ejects it at a higher pressure. In contrast to oil or mercury diffusion pumps, the working fluid (steam) is not condensed in the pump itself, but leaves it, moving in the same direction as the ejected (i.e. compressed) gas. Up to seven stages have been used in series and ultimate pressures of 10^{-1} torr and lower can be reached, the high-pressure stage ejecting straight to the atmosphere. Steam ejector pumps are particularly useful in large-scale industrial plant where steam and cooling water are already available, the pumping loads 'dirty', and oil vapour stream pumps therefore unsuitable. For details, including performance and economic considerations, see ref. 3.16.

§ 3.4. Chemical pumps

Chemical pumps are so called because the gas to be pumped forms a chemical (and generally irreversible) bond with the material of the pump. This distinguishes them from sorption pumps, which rely on Van der Waals bonds between gas and walls and from cryo pumps, which depend on bonds between like molecules. All three types of pump have the common property of retaining the pumped gas molecules, as opposed to expelling them.

3.4.1. *Sputter ion pumps*

In sputter ion pumps there is a cold-cathode discharge (see Appendix II, § 17) in the gas to be pumped, between an anode and

cathodes of a highly reactive metal, usually titanium. The impact
of the positive ions on the cathodes sputters (see Appendix II,
§ 18) the titanium on to surfaces of the pump, where the gas to
be pumped reacts with it and is immobilized. In addition, some
of the positive ions are driven into the cathode and there buried.
The latter action is especially important for the noble gases, which
do not react chemically with titanium.

Fig. 3·7 shows schematically the geometry of one simple design.

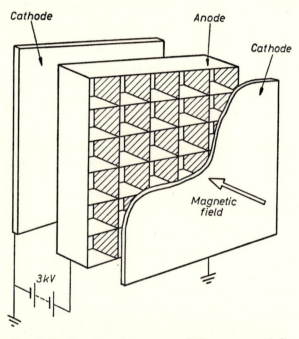

Fig. 3.7. Sputter ion pump [diagrammatic]
(Courtesy of Mullard Research Laboratories Ltd)

Electrons emitted from the cathodes oscillate in a potential well
between them produced by the positive potential of the anode, and
prior to an ionizing collision are constrained by a magnetic field
(order 1 to 2 kilogauss) to move in spiral paths, thus increasing the
ionization probability. The more massive positive ions, whose

motion is not seriously affected by the magnetic field, strike the cathodes and sputter titanium on to the walls of the anode cells.

Ions which are 'buried' in the cathode are liable to disinterment by subsequent ionic bombardment, and in sophisticated designs this possiblity is minimized by non-planar cathode geometrics (see ref. 3.6). The disinterment phenomenon is particularly trouble-some with argon (which constitutes about 1 per cent of atmospheric air) and leads (except in pumps specially designed to avoid it) to *argon instability* in which argon is periodically released from the cathodes causing steplike rises in pressure of as much as two decades. For a detailed account of sputter ion pumps, see Robinson (ref. 3.7).

Sputter ion pumps are particularly indicated when a system free from oil vapour is desired, hence they are usually pumped down to the starting pressure (10^{-2} torr or less) by a sorption pump (see § 3.5) or a trapped rotary pump. To get easy and reliable starting, certain precautions are necessary, the most important being to prepump to as low a pressure as practicable and to do so by means of an adequately sized connection to the starting pump. The pumping speed is a function of pressure and of previous history, but speeds between 10 litres sec^{-1} and 10000 litres sec^{-1} are available.

The advantages include low-power consumption, quietness (no mechanical backing pump needed), freedom from attention (no disastrous consequences of power failure), utility as its own leak detector (see § 5.10.4), and ability to produce, if required, pressures of 10^{-10} torr or less in an oil-free system.

Since the gas pumped is not ejected, the useful life is inversely proportional to the pressure, but at low pressure this is not a serious limitation, e.g. at 10^{-10} torr a 100 litres sec^{-1} pump will take about three years to saturate.

3.4.2. *Titanium sublimation pumps*

As a freshly deposited film of titanium chemically combines with nearly all gases except the noble ones and methane, any means of producing such a film on the inner surface of a vessel can form the basis of a pump. One such means involves a water-cooled anode supporting a pendant drop of titanium, which is heated to evaporation temperature by electron bombardment from a thermionic cathode, and continuously fed by a titanium wire. The evaporated titanium is condensed on water-cooled surfaces.

A simple method, with, however, limits to the pumping life, involves sublimation of titanium from an electrically heated filament of titanium/molybdenum alloy, and its condensation on water or L.N_2 cooled walls; the *sticking probability* of the gas molecules (i.e. chance of sticking to the Ti film on the first impact) being considerably greater on the cooler surface.

Sublimation pumps can be started at pressures of 10^{-3} torr or less, obtained by other means, and they are particularly useful when operated in short bursts to reduce pressure from say 10^{-8} to 10^{-11} torr.

Titanium sublimation pumps can have speeds reaching many thousands of litres per second, since, with suitable ('crab pot') geometry, the speed factor of the entry port (see § 3.3.4) can approach unity. The pros and cons of this type of pump have been well summarized by Power (ref. 3.8).

Small glass expendable sublimation pumps are available which consist of an ordinary Bayard-Alpert pressure gauge (see § 4.5.1), into which an extra getter titanium or zirconium-coated tungsten filament has been introduced. The system is evacuated to (say) 10^{-7} torr by conventional means, the getter filament then being heated and the pressure reduced to $\sim 10^{-10}$ torr. Prior to its use as a getter pump, the gauge can be used for pressure measurement in the ordinary way.

3.4.3. *Other chemical pumps*

A good getter material for a pump should be chemically active when freshly distilled, have a low vaporization rate at the maximum temperature at which vacuum vessels in which it is used is out-gassed, but evaporate readily at higher temperatures, and should not react unduly when exposed to the atmosphere.

Barium, alloyed with aluminium to produce a protective oxide film when exposed to atmosphere, is much used as a getter in electronic tubes. The mechanism of its uptake has been studied by Bloomer and Cox (ref. 3.9).

Numerous other metals have been used for special purposes, e.g. the pumping of H_2 and D_2 by evaporated Mo has been described by Hunt, Damm, and Popp (ref. 3.10).

A useful summary is given by Power (ref. 3.11).

§ 3.5. Sorption pumps

Sorption pumps rely on physical adsorption between *unlike* molecular species, viz. the adsorbent surface and the gas molecules to be pumped. The mechanism essentially consists in the formation of a monolayer* (or less), since in n-layer adsorption (where n exceeds 2 or 3) the vapour pressure of the adsorbed layer approximates to that of the adsorbent in bulk, and the action is more properly termed cryo pumping—to be discussed in § 3.6. The efficiency of sorption pumping is determined by a competition between the adsorption energy q, tending to hold a gas molecule on the adsorbent surface, and the thermal energy kT tending to dislodge it, the determining factor in the escape probability being $\exp(-q/kT)$ where k is Boltzmann's constant, and T the absolute temperature of the adsorbing surface (see Appendix II, § 19).

Hence, a good adsorbent should have a large accessible area per unit volume (implying porosity) and a value of q, such that $\exp(-q/kT)$ increases by a large factor when T is raised from a convenient low working temperature (e.g. L.N_2 at 77°K) at which the surface coverage is large, to the highest *activation* (i.e. degassing) temperature which does not permanently impair its low temperature adsorbent properties.

Two commonly used adsorbents are charcoal and artificial zeolites, the porosity of the former derives from the cellular structure of the original wood, and that of the latter from the molecular sized cavities left by dehydration, a process which does not involve the collapse of their original structure.

---

* See Appendix II, § 5.

The porosity of charcoals is such that, if the pores were cylindrical, they would be of diameter about 50 Å and, on this basis, the internal area of 1 cm³ of a good activated charcoal would be (ref. 3.12) about 10^3 metres².

The zeolites are three-dimensional Si–O–Al anionic networks, containing interstitial and exchangeable ions and neutral water molecules, the latter being removable by heating without structure collapse, leaving cavities of atomic dimensions. The size of the cavities varies from about 3 to 10 Å, depending on zeolite type. The zeolites adsorb selectively according to the relation between their cavity size and the size of the molecule to be sorbed, hence the name *molecular sieves*. The *equivalent internal surface* of the synthetic zeolites is about 600 to 800 metres² gram^{-1}.

Zeolites are activated (i.e. freed from water and other adsorbates) by heating to several hundred degrees centigrade *in vacuo* or by flushing, when heated, with dry nitrogen. When activated under optimum conditions, one gram of a typical zeolite (Molecular Sieve 5A manufactured by Union Carbide) will, when in equilibrium with 10^{-3} torr in the gas phase, carry a burden of about 100 cm³ (measured at N.T.P.) of nitrogen at 77°K (ref. 3.13), which is several orders greater than the burden under the same pressure at room temperature; but the pre-adsorption of 5 per cent of water by weight will reduce the burden it will carry at 10^{-3} torr and 77°K to about 20 cm³ measured at N.T.P. At 77°K the zeolites adsorb useful amounts of most common gases. H_2 is little adsorbed, and Ne and He hardly at all. The presence of the latter two noble gases is a limiting factor in the vacua which may be obtained starting from air at atmospheric pressure. The adsorption of zeolites increases significantly with the number of previous sorption and reactivation cycles they have undergone, but is very adversely affected by pre-adsorption of water, which is one of the most difficult of the common gases to remove. If the adsorption of water is avoided, reactivation necessary for effective pumping of dry air only involves letting the pump warm up to room temperature after use at 77°K.

The zeolites are usually in the form of pellets, of millimetre dimensions, held together with 20 per cent of clay. Since the thermal conductivity of zeolites is small, and since as soon as they produce vacua such that the mean free path is limited by inter-pellet distance the effective conductivity decreases, the design should be such that the maximum distance of any part of the zeolite from the refrigerant-cooled surface does not exceed (say) 1 cm.

The metal gauze cylinder shown in Fig. 3.8 facilitates access of

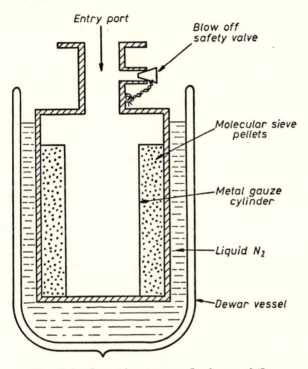

Fig. 3.8. Sorption pump [schematic]

the gas to be pumped to the zeolite-filled adsorbing annular space. The safety valve vents to atmosphere the gas desorbed when the pump returns to room temperature.

Precise quotation of pumping speeds is not possible, but for easily adsorbed gases like O_2, N_2, and Ar, the speed of a well-designed pump is often of the order of the conductance of its neck, implying that the chance that a molecule which enters the pump will escape sorption is small.

Though most commercially available pumps are designed for refrigeration by L.N$_2$, lower temperatures are required for sorption pumping of H$_2$, Ne, and He, and pumps have been designed and tested (refs. 3.14 and 3.15) using molecular sieves cooled with liquid H$_2$ (b.p. 20·3°K) or liquid He (b.p. 4·2°K). It is believed that such pumps may be useful in evacuating large space-simulation chambers to very low pressures.

An important common use of L.N$_2$-cooled sorption pumps is to provide the starting pressure (10^{-2} torr or less) for sputter ion pumps in systems where complete freedom from the possibility of organic contamination is desired (see § 3.41).*

§ 3.6. Cryo pumps

In cryo pumps the gas to be pumped is condensed on the solid phase of gas already pumped, so that a thickness of many molecular layers may be built up. For high efficiency, it is therefore necessary that the sticking coefficient of impinging gas molecules on already condensed gas should be high and that the solid so built up should have a saturated vapour pressure smaller than the ultimate pressure the pump is desired to achieve.

The simplest cryo pump is the L.N$_2$-cooled vapour trap which is very efficient in pumping water, whose vapour pressure at 77°K is completely insignificant. For this reason the addition of a L.N$_2$-cooled trap to an oil diffusion pump may be justified in situations where the advantage of its excellent speed for H$_2$O outweighs the disadvantage of decreased speed for permanent gases, due to decreased conductance to the diffusion pump.

However at 77°K many common gases are not condensed, so that cryo pumps are usually cooled by L.H$_2$ (b.p. 20·3°K) or L.He^4 (b.p. 4·2°K). To pump H$_2$ to pressures below about 10^{-6} torr a condensing surface cooler than 4·2°K is required, these can be

---

* A work chamber of e.g. 20 l volume can be quickly and cheaply pumped to ~10^{-4} torr (mainly He and Ne) following use of a L.N$_2$ trapped water jet pump.

obtained, for instance, by boiling He[4] at reduced pressure, e.g. at about 60 torr, when its b.p. is 2·3°K at which temperature the vapour pressure of H_2 is about 10^{-15} torr.

To minimize consumption of L.H_2 or L.He and to prevent excessive temperature rise of the surface of the condensed layer of gas above that of the underlying cryo panel, radiation shields, often cooled with L.N_2, are arranged so that the cryo surfaces do not 'see', and therefore receive radiation from, the surrounding world at about 300°K.

The construction is shown schematically in Fig. 3.9. The condensing cryo panel, cooled by L.He, is surrounded by a

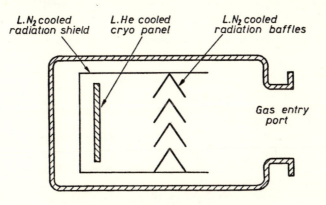

Fig. 3.9. Cryo pump [schematic]

radiation shield cooled by L.N_2, the gases to be pumped reaching the cryo panel via cooled baffles which limit entry of unwanted radiation without unduly decreasing pumping speed.

Cryo pumping is usually preceded by evacuation via mechanical or diffusion pumps to pressures of order 10^{-2} torr both to economize in refrigerant (since condensation of gas liberates latent heat) and to increase the useful life of the pump before the layer of condensed gas becomes so thick, and its thermal resistance therefore so large, that the condensing surface becomes significantly warmer than the underlying cooled cryo panel.

Cryo pumps have pumping speeds high in relation to their

overall dimensions. For example, in one type the pumping surface is a L.He⁴-filled vertical cylinder, surrounded by a L.N₂-cooled co-axial cylinder containing chevron baffles, the assembly being located within the work chamber itself. The approximate overall dimensions of the outer chevron-baffled cylinder which constitutes the radiation shield are* 22 cm (height) by 13 cm (diameter) and the pumping speed is 2000 litres sec⁻¹ for air and 6000 litres sec⁻¹ for H_2.

Cryo pumps are used in a number of special situations in particular for evacuating large low-density wind tunnels and large space-simulation chambers. In small laboratory apparatus a 'finger' cooled by liquid helium is useful in quickly obtaining a good vacuum, e.g. in experiments on clean surfaces.

* Data supplied by courtesy of Balzers AG, Liechtenstein.

References

3.1. POWER, B. D., 1966, *High Vacuum Pumping Equipment*, pp. 21 *et seq.* (Chapman & Hall Ltd).

3.2. POWER, B. D., 1966, *High Vacuum Pumping Equipment*, Chapter 5 (Chapman & Hall Ltd).

3.3. POWER, B. D., 1966, *High Vacuum Pumping Equipment*, pp. 190 *et seq.* (Chapman & Hall Ltd).

3.4. BECKER, W., 1966, *Vacuum*, 16, pp. 625 *et seq.*

3.5. POWER, B. D., 1966, *High Vacuum Pumping Equipment*, pp. 55 *et seq.* (Chapman & Hall Ltd).

3.6. LEWIN, G., 1965, *Fundamentals of Vacuum Science and Technology*, p. 156 (McGraw-Hill Book Co.).

3.7. ROBINSON, N. W., 1968, *The Physical Principles of Ultra-High Vacuum*, pp. 50 *et seq.* (Chapman & Hall Ltd).

3.8. POWER, B. D., 1966, *High Vacuum Pumping Equipment*, pp. 310 *et seq.* (Chapman & Hall Ltd).

3.9. BLOOMER, R. N., and COX, B. M., 1965, *Brit. J. App. Phys.*, 16, p. 1331.

3.10. HUNT, A. L., DAMM, C. C., and POPP, E. C., 1961, *J. App. Phys.*, 32, pp. 1937–1941

3.11. POWER, B. D., 1966, *High Vacuum Pumping Equipment*, pp. 287 *et seq.* (Chapman & Hall Ltd).

3.12. DUSHMAN, S., Edited by LAFFERTY, J. M., *Scientific Foundations of Vacuum Technique*, 2nd Edition, 1962, p. 437 (John Wiley & Sons Inc.).

3.13. STERN, S. A., and DiPAOLO, F. S., 1967, *Jour. of Vac. Sci. and Tech.*, 4, pp. 347–355.

3.14. STERN, S. A. *et al.*, 1965, *Jour. of Vac. Sci. and Tech.*, 2, pp. 165–177.

3.15. STERN, S. A. *et al.*, 1966, *Jour. of Vac. Sci. and Tech.*, 3 99.

3.16. BECK, A. H., 1964, Editor, *Handbook of Vacuum Physics*, Vol. 1, pp. 77 *et seq.* (Pergamon Press Ltd).

3.17. POWER, B. D., and CRAWLEY, D. J., 1954, *Vacuum*, 4, pp. 415–437.

4

MEASUREMENT
OF
PRESSURE

§ 4.1. Introduction

The state of affairs in a vacuum system is determined by the molecular species present, the number density n (number per unit volume) of each, and the temperature, which in an isothermal situation will be the same for each species.

The total pressure is the sum of the partial pressures of the constituents, and since

$$P = nkT \qquad (4.1)$$

for each species (see Appendix I, § 5) the total pressure P is determined by the temperature, and the number density of the constituents.

The pressure (in sense of force per unit area) is seldom important in vacuum technology, the interesting quality usually being the number density n—a quantity measured by nearly all common gauges, except the McLeod, which despite grave limitations is the standard calibrating instrument for many gauges.

This chapter will deal first with the McLeod and its limitations, with some other common gauges and their calibration, with thermal transpiration, very briefly with mass spectrometers considered as instruments for determining relative number densities of various constituents, and with sorption/desorption methods.

In technology, it is often sufficient that the reading of the pressure gauge does not exceed a value known from previous experience to be acceptable, and so the inaccuracy of gauge indications is not serious. In some scientific work, however, e.g. investigation of chemical gas kinetics at low pressures, the exact numerical value of the number density may be desired.

§ 4.2. Survey of types of gauge measuring total pressure

(a) Barometric types

The liquid may be mercury or oil. Without instrumental aid, the minimum difference of level which can be detected is of order 10^{-1} mm, so mercury types have a lower useful limit of about 1 torr if 10 per cent accuracy is required, the sensitivity of oil types being about 15 times greater. In some situations, the considerable improvement of sensitivity afforded by instrumentation (optical or electronic) may be justified.

(b) Mechanical types

The most usual type is the diaphragm gauge in which the pressure causes mechanical movement of a septum forming one wall of an evacuated enclosure. The principle is identical with that of the aneroid barometer. A useful commercially available form has a circular scale covering 0 to 20 torr. The sensitivity of gauges of this type may be considerably increased by suitable instrumentation.

(c) Gauges of the McLeod type

The basic principle is to take a sample of the gas, decrease its volume by a known factor and measure the resultant pressure. The manometric liquid is usually mercury; details are given in § 4.3. It assumes Boyle's Law, i.e. PV is constant at constant T.

(d) Thermal conductivity gauges

These depend on the decrease of thermal conductivity at low pressures (see Appendix I, § 10) resulting in increase of thermal insulation of a heated body as pressure is reduced. Details are discussed in § 4.4.

(e) Hot-cathode ionization gauges

In these the number density is deduced from the amount of ionization produced by the passage of a known number of electrons of fixed initial energy through a fixed geometrical structure. Details are given in § 4.5.

(f) Cold-cathode ionization gauges

In these, the pressure is deduced from the current in a cold cathode discharge tube under controlled conditions of applied voltage. Details of the Penning gauge, which utilizes this principle are given in § 4.6.1. Useful qualitative indications are given by an H.F. discharge as mentioned in § 4.6.2.

(g) Gauges depending on other physical properties

These include gauges utilizing the viscosity decrease at low pressures (analogous to the decrease of thermal conductivity utilized in the thermal conductivity gauges), and the Knudsen gauge, which depends on the recoil force on a warm surface when a molecule rebounds from it (see §§ 4.7 and 4.8).

The most common gauges are mechanical gauges of the diaphragm type, thermal conductivity gauges, hot cathode gauges, and cold cathode gauges, while the McLeod is much used for calibration.

§ 4.3. The McLeod gauge

The general principle of operation, stated in § 4.2(c), viz. compression of a sample of gas by a known factor and measurement of the resultant pressure, is valid only if no significant constituent of the gaseous mixture is caused by the compression to condense to liquid or be adsorbed by the walls of the gauge.

Figs. 4.1(a) and (b) show alternative methods of operation, the geometry of the glass tubing being shown for simplicity, as identical. An auxiliary vacuum is applied to reservoir R, so that the mercury level falls below P and the bulb B is filled with gas at the pressure p_0 to be measured. Atmospheric air is then slowly

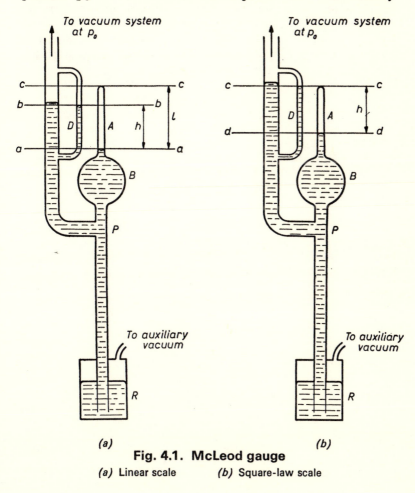

Fig. 4.1. McLeod gauge

(a) Linear scale   (b) Square-law scale

admitted to R, so that the mercury rises and, on passing P, traps a sample of gas in the volume (V) between P and the closed end of the capillary (cross section area S) which is sealed to the bulb B. The subsequent procedure differs as between the two modes.

In the mode of Fig. 4.1(a), the mercury is allowed to rise until it reaches a fixed mark at level a–a. The gas compression ratio is then V/lS and its pressure is therefore $p_0(V/lS)$, where l is the distance in millimetres between a–a and the level c–c of the top of the capillary tube. The level b–b to which the mercury rises in the limb D (which, to minimize uncertainties due to surface tension, is of the same bore as A) is noted. If h is the vertical distance in millimetres between b–b and a–a, the total pressure at a–a is $p_0 + h$ mm of mercury, i.e. $p_0 + h$ torr. Hence

$$p_0 + h = p_0 \frac{V}{lS}$$

or
$$p_0 = \frac{h}{V/lS - 1} \qquad (4.2)$$

i.e.
$$p_0 \approx \frac{lSh}{V} \text{ millimetres of mercury (torr)}^* \qquad (4.3)$$

since in practice $V \gg Sl$. Thus a scale fixed behind D, from which the height h is measured, is linear in p_0.

When operated as shown in Fig. 4.1(b), air is admitted to R until the mercury in D has reached the level c–c of the tip of the capillary c and the distance h millimetres between c–c and d–d is read. The compression ratio is V/Sh and the pressure p_0 is given by

$$p_0 + h = p_0 \frac{V}{Sh}$$

$$\therefore p_0 = \frac{h}{V/Sh - 1} \qquad (4.4)$$

or
$$p_0 \approx \frac{Sh^2}{V} \qquad (4.5)$$

since
$$\frac{V}{Sh} \gg 1$$

* With h and l in mm, S and V must be in mm² and mm³ respectively.

Thus the pressure, which is read on a scale fixed behind A, is proportional to the square of h.

The advantage of the mode shown in Fig. 4.1(a) is linearity of scale, so interpolation by eye between scale markings is easy. On the other hand, the compression ratio is fixed, while in the mode of Fig. 4.1(b) it is variable, increasing as the pressure becomes lower.

The practicable compression ratio (i.e. pressure amplification) is determined by (i) the maximum weight of mercury which can conveniently be supported in a glass bulb (the volume V) and (ii) the minimum allowable value of S, which increases with the surface tension depression and 'stiction' of mercury meniscus in D and A, and varies with the cleanness of the mercury and the surface state of the glass. These factors may not be the same in D and A, even though they are nominally identical, and thus become serious at small diameters. The advantage of a compression ratio which increases as pressure decreases, and the consequent improvement of sensitivity at low pressure and larger useful working scale, has made the 'square law' McLeod the more popular instrument.

To get some 'feel' for magnitudes, consider a square-law gauge with $V = 10^2$ cm^3 (corresponding to nearly 1½ kg of mercury), $S = 10^{-2}$ cm^2 (i.e. of capillary diameter of order 1 mm) used to measure a pressure of 10^{-3} torr. Then from equation (4.5), $h = 10$ mm, and if the error in reading h is 0·5 mm of scale length, the error in p is about 10 per cent.

The above example is not intended to suggest that considerably better accuracies cannot be achieved, by using larger V and small S. However, for good accuracy, special care must be taken, especially in the surface treatment of the inner surface of the capillary and the uniformity of its bore. It is probably reasonable to assume that unless special precautions are taken, the gauge is not reliable as an absolute standard below about 10^{-4} torr. Leck (ref. 4.1) should be consulted for details.

4.3.1. The mercury-vapour stream effect

Unless isolated by an intervening cold trap, a McLeod gauge

will contaminate with mercury vapour the vacuum system to which it is connected, to a degree dependent on the temperature and therefore vapour pressure of the mercury. But the cold trap, in protecting the system from mercury vapour, introduces another error. The vapour coming from the McLeod, and condensing in the trap, produces a pumping effect causing the gas pressure indicated by the gauge to be less than that in the system on the side of the trap remote from the gauge.

The magnitude of the effect increases with the diameter of the pipe connecting gauge and cold trap, and with the temperature and therefore vapour pressure of the mercury. Since the inequality of gas pressure set up by the stream of mercury vapour moving from gauge to trap is opposed by the diffusion of gas in the opposite direction, the effect is less the greater the diffusion coefficient of the gas through mercury vapour. The resultant error for the rare gases and for N_2 and CO_2 has recently been measured by Elliott, Woodman, and Dadson (ref. 4.2), whose paper includes useful references to the literature. They state that the effect is proportional to the tube diameter and, with a tube of effective diameter 1 cm and a temperature of 23°C, is about 2 per cent for He and 27 per cent for Xe over the pressure range of 4×10^{-6} to 1×10^{-4} torr. The error can be reduced by refrigerating the McLeod (ref. 4.3) (Hg remains liquid down to $-38\cdot8$°C) and by reducing the bore of the tube between gauge and cold trap. For necessary precautions in the latter case see ref. 4.4.

§ 4.4. Thermal conductivity gauges

In general the heat loss from a hot body placed in a vessel with cooler walls depends on geometry and the thermal conductivity of the gas. However, as the pressure is reduced, a stage is reached at which the thermal conductivity decreases with pressure and indeed the term itself loses its usual meaning. Consider a hot wire centrally placed in a vessel with walls at room temperature. At sufficiently high pressure, there will be a temperature gradient from wire to wall and the situation may be expressed in terms of geometry and a constant thermal conductivity; see e.g. ref. 4.5. When the pressure is so low that the mean free path is large compared with the wire diameter, a molecule which has hit the wire and been heated, will subsequently make many collisions with the other molecules and with the walls of the vessel before again hitting the wire. Hence the temperature of molecules hitting the wire will be approximately that of the walls. In these circumstances, the rate of heat loss decreases linearly with bombardment rate, i.e. pressure.

This mechanism governs the behaviour of the three common types of thermal conductivity gauge—Pirani, thermocouple, and thermistor—now to be described.

4.4.1. *The Pirani gauge*

The basic principle of this gauge is that the heat loss from a current-carrying wire, diameter d, mounted in an envelope connected to the work chamber decreases with pressure if $\lambda \gg d$, where λ (see Appendix I, §§ 9 and 10) is the mean free path.

The principle has been applied in two ways: (*a*) keeping the temperature of the wire constant as the pressure changes, the necessary voltage change being a measure of change of pressure, or (*b*) allowing the temperature of the wire to change as the pressure changes, and observing the resistance change of the wire.

Method (*b*) is more common, the wire being one arm of a Wheatstone bridge, and the pressure change being indicated by change in the out-of-balance current in the bridge micro-ammeter.

Fig. 4.2 shows the essentials of a constant-voltage bridge, the arm R_1 being a tungsten wire in a glass envelope connected to the work chamber whose pressure it is desired to read on M. To zero the instrument the pressure in the envelope of R_1 is reduced to a value such that the heat loss *via gas* is negligible compared with that via radiation and thermal conduction from the tungsten to its leads, the potentiometer R_5 being adjusted to make the reading of the meter M zero. If gas is now introduced into the envelope of R_1 the tungsten filament is cooled, the bridge unbalanced and a meter reading produced which is a function of p.

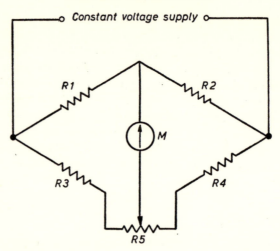

Fig. 4.2. Constant-voltage Pirani bridge

To compensate for changes of ambient temperature the temperature coefficient of R_2 is chosen to be approximately equal to that of R_1, though as Leck points out (ref. 4.6) the compensation is only effective over a limited pressure range. At a given gas temperature and pressure, the rate at which the molecules bombard the filament is proportional to (molecular weight)$^{-1/2}$ (see Appendix I, § 6) and the amount of heat carried away per impinging molecule depends on the difference between temperatures of wire and gas, the specific heat of the gas, and its thermal accommodation coefficient α (see Appendix II, § 6) on the tungsten filament.*

Thus the calibration of a Pirani, usually done for N_2 or air, depends on the gas. For example, a gauge calibrated for N_2 will read about 60 per cent high for H_2 at the same pressure, because H_2 is the more efficient coolant.

A useful arrangement which doubles the sensitivity is to make all four arms of tungsten, R_1 and R_4 being in a single bulb which is at the pressure p to be measured, R_2 and R_3 being mounted in a

---

* Fortunately, in the relatively high-pressure region in which Piranis are used, adsorbed gases make α approximately constant and not far from unity.

similar bulb which is evacuated and permanently sealed off. The two bulbs are put close together, preferably in a thermally insulating enclosure, to compensate for changes in ambient temperature. Numerical magnitudes of a typical arrangement are as follows. R_1, R_2, R_3, and R_4 are tungsten filaments of about 0·05 mm diameter, with a room temperature resistance of order 20 Ω, which attain a temperature of about 300°C when a stabilized 2-volt supply is applied to the bridge. A 0–200 microammeter, resistance of order 200Ω, gives a full-scale deflexion corresponding to a N_2 pressure of 4×10^{-3} torr. The minimum useful reading is 10^{-4} torr. By shunting the meter, three or more ranges going up to (say) 10^{-1} torr can be obtained with this arrangement, but linearity is sacrificed at the higher pressures, the instrument becoming increasingly insensitive.

However, by special design (ref. 4.7) instruments with a useful range of up to 1 torr are possible, but above this pressure other types of gauge, e.g. the diaphragm gauge, are usually preferable. It is possible to construct considerably more sensitive Piranis, but these are subject to excessive zero drift due to causes such as changes in ambient temperature and in radiation and gas thermal accommodation surface properties of the filaments.

By deliberate exploitation of gas convection as opposed to conduction thermal conductivity gauges may be used up to atmospheric pressure; see e.g. Steckelmacher, who give a useful account of pressure measurement from an industrial viewpoint (ref. 4.8).

The speed of response of a constant-voltage Pirani gauge is determined by the thermal capacity of the W filament, and degree of thermal linkage between filament and gas; the time constant is usually of the order of some seconds. However, by using a filament of diameter 10^{-2} mm, maintained at a constant resistance by feedback circuits, a time constant can be attained which is 0·14 seconds at 10^{-6} torr and 0·02 seconds at 10^{-1} torr (ref. 4.9).

4.4.2. *The thermocouple gauge*

The principle, identical with that of the Pirani, is the variation of heat loss from an electrically heated wire in the pressure region where it decreases with decrease of gas pressure.

One simple form consists of a platinum ribbon heated by constant current to between 100 and 200°C, the temperature, which is a function of pressure, being measured by a thermo-couple attached to it, and the output read on a sensitive micro-ammeter.

The scale is non-linear and the lower limit of use is about 10^{-3} torr. Most commercial instruments cover the range 10^{-3} to 1 torr, but other ranges can be covered by appropriate design.

The advantages of this gauge are ruggedness, simplicity and as in the Pirani gauge the use of a filament running at a low tempera-ture, which is therefore not injured by exposure to atmospheric air. It is also useful in detecting sudden and accidental pressure rises and activating the necessary 'shut down' devices. Like the Pirani, its calibration is a function of gas species.

4.4.3. *The thermistor gauge*

This depends upon the heat loss from a small bead of semi-conducting material fastened to two leads, and enclosed in a small bulb connected by tubulation to the vacuum work chamber. The material has a large negative temperature coefficient of resistance, and is self-heating when a current is passed through it. Increase in gas pressure causes increase in cooling, decrease of temperature, and increase of resistance and of voltage drop (ref. 4.10)

One commercial form has two scales, viz 0–100 microns and 0 to 760 torr. The scale of the former is linear in pressure, but the latter is much less sensitive, the scale reading being more nearly proportional to the logarithm of the pressure. The calibration, as in the other thermal conductivity gauges, depends on the gas, the reading being for example higher for H_2 than for N_2 at the same pressure. The extension of the upper limit of measurement of a thermistor gauge, by surrounding the thermistor bead with an electrically non-conducting powder, is described by Green and Lee (ref. 4.11).

§ *4.5. Hot-cathode ionization gauges*

These gauges depend (see Appendix II, §§ 11 and 12) on the thermionic emission from a hot filament of electrons, which are accelerated by an electric potential to a velocity which maximizes their probability of ionizing the gas molecules which they hit. The positive ions so produced are collected by an electrode whose potential is such that it cannot collect electrons, which if collected would diminish the current attributed to the positive ions.

The rate of ion production is proportional to the number density of the gas molecules and to the number of ionizing electrons emitted per second from the hot filament. Hence, for a given ionizing electron current i_g, the positive ion current produced i_c is proportional to the number density of gas molecules n, which is related to the pressure P by the relation

$$P = nkt \tag{4.6}$$

The characteristics of an ionization gauge are therefore expressed by the equation

$$i_c = Gi_gP \tag{4.7}$$

where G is called the gauge sensitivity, and is specified as amps per amp per torr, i.e. the sensitivity is given in $torr^{-1}$. Gauge sensitivities are usually quoted for N_2, room temperature being implied; the specification of temperature is relevant, since the gauge really measures n, and not P.

The accelerating voltage, applied to the electrons from the filament, is usually between 100 and 200 volts for the reason given in Appendix II, § 11.

The underlying assumptions and range of validity of equation 4.7 are discussed in ref. 4.12.

The use of a hot-cathode ionization gauge to measure gas pressure involves the possibility of several undesired side effects. One effect (which causes the gauge to read high) is the emission of adsorbed gas from its electrodes under the influence of radiation and/or electrons from the filament. The remedy is to outgas all parts of the gauge at temperatures higher than will be attained in service (ref. 4.13). An effect which may cause too low a gauge reading is the reaction of the hot filament with the gas. For instance O_2 reacts with hot W producing an oxide which is deposited on the gauge walls (ref. 4.14). Also positive ions or molecules excited by electron impact may react with the walls of the gauge (see e.g. ref. 4.15) which become in effect a pump.

All these effects are minimized by using a filament of minimum area, run at as low a temperature as possible, using the smallest practicable value of i_g, and connecting the gauge to the work chamber by tubulation of as large a bore as possible. The limiting case of the latter is to mount the gauge element in the work chamber itself, the so called 'nude gauge' technique.

4.5.1. The Bayard-Alpert gauge

Fig. 4.3 shows the essential figures of a B-A gauge, the thermally emitted electrons from the hot filament (usually W) being accelerated to the grid; they make a few oscillations through its wires and are finally captured by it, constituting the i_g of equation 4.7. Before their final capture, a small fraction of these electrons ionize gas molecules, and most of the ions produced inside the grid are collected by the thin wire in its centre and constitute the current i_c. This current, amplified externally as necessary, is for a given value of i_g, a direct measure of the number density n, or gas pressure P. Stable sensitivity demands stable potentials within the gauge, so it is advantageous to control the potential of the inner surface of the glass envelope by means of a thin conducting film. Cobic, Carter, and Leck (ref. 4.16) have investigated the occurrence of a bistable mode of operation of a B-A gauge, due to envelope potential.

The electrons from the filament, on striking the grid, generate soft X-rays (see Appendix II, § 15) which in turn cause the emission of photoelectrons from the collector, thus increasing the *apparent* current i_c, and to that extent falsifying the readings of the gauge. In the B-A gauge, this soft X-ray effect does not become serious (i.e. comparable with the ion current) until the pressure falls

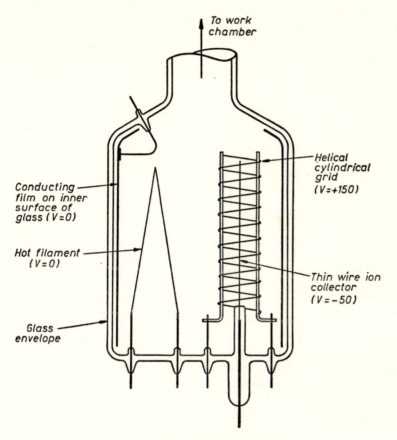

Fig. 4.3. B-A ionization gauge

(Figures in brackets are volts with respect to filament)

to about 10^{-10} torr; the collector is a *thin* wire (and hence has small target area for photons) in order to achieve this. Earlier types of ionization gauge, in which the electrode arrangement is different and the collector much larger, have a lower limit of pressure of about 10^{-7} torr, due to soft X-ray effect. The sensitivity G of the B-A gauge is, like that of the older types, between 10 and 20 $torr^{-1}$.

Typical magnitudes are as follows:

Filament emission to grid (i_g) = 100 micro amps
$$G = 10 \text{ torr}^{-1}$$
$$\text{pressure} = 10^{-8} \text{ torr}$$

Hence, the ion current (i_c) is 10^{-5} micro amps. So, except for the higher pressures, an amplifier of the ion current is necessary.

The upper limit of the B-A gauge is about 10^{-3} torr, owing to loss of linearity and to the fact that the hot filament has in general a very short life at pressures greater than this. But thoria-coated iridium filaments have a longer life in 'gassy' conditions.

4.5.2. *Other hot-cathode gauges*

Ways of improving the performance of hot-cathode gauges include (a) using filaments with good electron emission at low filament temperature (ref. 4.17), thus reducing gas/filament chemical reaction, (b) employing a geometry of electric fields which increases the number of ions per emitted electron, hence both reducing the relative importance of the soft X-ray effect and requiring a lower filament temperature and so reducing gas/filament reaction rate, (c) reducing the soft X-ray effect by arranging (ref. 4.18) that the ion collector cannot 'see' any surfaces bombarded by electrons.

Chapter 6 contains further notes on gauges specially designed to minimize spurious residual collector currents, which are a major difficulty in measuring pressures below 10^{-10} torr. However, because of its high sensitivity due to very long electron paths, achieved without use of a magnetic field (often inconvenient and sometimes unacceptable), the Orbitron gauge is mentioned here. It consists of a short thermionic filament located off axis in a cylindrical positive ion collector, down the axis of which is a wire anode. The ion collector is grounded, the filament is about 10 volts positive to it, and the central anode is a few hundred volts positive to the filament. The geometry is such that some of the electrons have an angular momentum with respect to the anode, and therefore long spiral paths round the anode before hitting it—typical length is of order 1 metre. In fact, they orbit the anode rather as a satellite orbits a planet. The electron current necessary to produce a given ion current and the soft X-ray effect are therefore very low and the sensitivity very high—values for sensitivity G of order 10^3 torr^{-1} being quoted (refs. 4.19 and 4.20).

The problems particular to measuring very low pressures stem from the soft X-ray effect and other residual effects, which become dominant over ionization of gas molecules, rather than from difficulties in measuring the collector current itself. This will be further discussed (§ 6.2) in relation to ultra-high-vacuum techniques.

§ 4.6. Cold-cathode ionization gauges

Cold cathode gauges are immune from a major hazard in hot-cathode gauges (i.e. filament burn-out caused by too high a pressure), but in most types it is necessary to provide a magnetic field and hence a permanent magnet to sustain the discharge at low pressures. Also the voltages required are, in general, higher than those of thermionic filament gauges, and the unwanted pumping effect is greater.

4.6.1. *The Penning cold-cathode gauge*

This consists (see Fig. 4.4) of a cylindrical anode at about +2 kV between two flat-disc cathodes at ground potential placed between the poles of a permanent magnet which gives a field of about $\frac{1}{2}$ kilogauss. As the pressure is reduced, a discharge is struck at about 10^{-2} torr and the current, which is the sum of the positive ions reaching the cathodes and the secondary electrons liberated from it, decreases with pressure till it goes out at about 10^{-6} torr. The electrons are constrained by the magnetic field to move in spiral paths in the potential trough between the two cathodes (see Appendix II, § 16), thereby increasing their free path before capture by the anode. Thus the chance of producing positive ions is enhanced and the discharge is maintained to much lower pressures than would otherwise be the case.

5

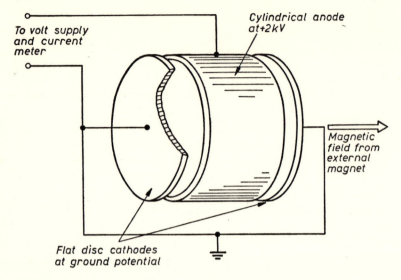

To volt supply
and current
meter

Cylindrical anode
at +2 kV

Magnetic
field from
external
magnet

Flat disc cathodes
at ground potential

Fig. 4.4. Elements of Penning cold-cathode ionization gauge

The sensitivity is of order 10 amps per torr for N_2, so that the discharge current can be read on a microammeter. A typical two-range commercial instrument would cover 10^{-2} to 10^{-4} and 10^{-4} to 10^{-6} torr. The merit of the gauge is its ruggedness; its disadvantages are that it is liable to sudden fluctuations of current and therefore of calibration (ref. 4.21) and its cathodes are sputtered under the positive ion bombardment, causing pumping (which may be between 0·1 and 1·0 litres sec^{-1}) and impairing the insulation of the anode.

4.6.2. *Other cold-cathode gauges*

These devices, which are of magnetron geometry, have cylindrical symmetry and orthogonal electric and magnetic fields, the electrical field being radial, the anode and cathode coaxial cylinders, and the magnetic field axial. The magnetic field bends the paths of the electrons (see Appendix II, § 16, paragraph 1) so that they tend to return to the cathode whence they came unless they lose velocity by ionizing a gas molecule and so decrease the restraining effect of the magnetic field. Further details of cold-cathode gauges designed specifically for measurement of very low pressures are given in Chap. 6.

The Tesla coil, the high-frequency oscillations of which can excite electrodeless discharges in gases over a limited pressure range, cannot strictly be described as a cold-cathode gauge, but is useful as a qualitative pressure indicator over the pressure range 10 torr to about 0·05 torr. A typical instrument generates oscillations of 4–5 MHz frequency, and will, if held near the glass part of a vacuum system, produce luminous effects in the gas, the appearance of which is a very rough indication of pressure. It can be applied (hand-held) to any desired glass part of the system and immediately gives some information if the pressure is in its working range. If no discharge is seen the pressure is either many torr or below 10^{-2} torr. The colour of the discharge is an indication of the nature of the gas being excited (ref. 4.22).

§ 4.7. The Knudsen gauge

The Kundsen gauge may be classed as both mechanical and thermal since it depends on the momentum change thermally produced, when a cold molecule rebounds from a warmer surface having been heated thereby. It can be designed for use at low pressures (e.g. 10^{-4} to 10^{-7} torr), has a calibration independent of gaseous species, except in so far as they affect thermal accommodation (see Appendix II, § 6) on the warm surface, and does not need a hot filament, with its attendant problems of interaction with the gas. In some cases, these factors outweigh the disadvantage of its complexity and fragility. Ref. 4.23 gives an excellent account of this gauge which will not be further described here.

§ 4.8. The viscosity gauge

This gauge is of two types. In one—the oscillating vane type—a surface initially set in oscillatory motion is damped by molecules which impinge upon it, partake of some of its energy, and rebound. It is usable in the range 10^{-2} to 10^{-5} torr. It is claimed that, at a given temperature and pressure, the decrement of its oscillation is proportional only to the square root of the molecular weight of the gas, and hence the calibration, determined for one gas, is known for all.

The validity of this statement implies the questionable assumption that the degree of accommodation of energy and momentum on a surface is the same for all gaseous species. Mair (ref. 4.24) has recently reviewed the consequences to the theory of oscillating gauges of various assumptions regarding accommodation and angular distribution of rebounding molecules.

An outstanding advantage of the oscillating vane type is that it can be constructed entirely of fused silica, and hence can be used for work on corrosive gases and vapours. However, getting a reading takes time, which at 10^{-4} torr may be of order 10^2 seconds.

The other type of viscosity gauge is the rotating type, in which a surface is kept in continuous rotation, its plane being parallel to a similar surface whose rotation is elastically constrained. Molecules strike the rotating surface and rebound, having acquired an additional component of momentum parallel to the direction of movement. Part of this is transferred to the elastically constrained member, causing it to rotate to an equilibrium position which is a function of pressure. The gauge is capable of measuring pressures as low as 10^{-6} to 10^{-7} torr.

An account of the theory and practice of viscosity gauges is given in refs. 4.25 and 4.26, and as these gauges are not widely used they will not be further discussed here.

§ 4.9. Gauge calibration

The three methods commonly used are (*a*) comparison with a McLeod gauge, (*b*) the reduction, assuming Boyle's Law, of the pressure of a known small volume of gas at a known (e.g. atmospheric) pressure by a known expansion, (*c*) a dynamic method, depending on the pressure drop across an aperture of known conductance. Meinke and Reich have compared methods (*b*) and (*c*) for He and Ar over the range 10^{-2} to 10^{-6} torr (ref. 4.27).

4.9.1. *Comparison with a McLeod gauge*

This gauge, already described in § 4.3 can, if proper precautions are taken, be used as a standard to calibrate B-A ionization gauges down to pressures approaching 10^{-6} torr (ref. 4.2), with an accuracy of 1 or 2 per cent, but only for non-condensible gases. Since it employs mercury, an associated cold trap with its consequent correction for the vapour stream effect (see 4.3.1) is necessary.

It is, however, widely used and its sources of error and the precautions to minimize them have been extensively studied.

4.9.2. *Gas-expansion method*

In this technique (originally due to Knudsen) a small known volume V_1 of calibrating gas is expanded into a known much larger volume V_2, a small portion of which (V_3) can be isolated. The remainder of the gas is pumped away and the gas in V_3 again expanded into a large volume which may conveniently be V_2. The process is continued as necessary and the final pressure calculated by Boyle's law. Two possible sources of error are (*a*) the outgassing of the apparatus which would add unknown amounts of gas, and therefore pressure to the apparatus and (*b*) adsorption causing disappearance of gas. Effect (*a*) can be minimized by suitable pretreatment, but effect (*b*) is in principle unavoidable, though in practice it is apparently not significant with permanent gases such as N_2 and Ne (ref. 4.2).

4.9.3. *The dynamic method*

This is a steady-state method, in which a known and constant influx of gas $\dot{Q}$, conveniently measured at or near atmospheric pressure, enters the upper part of a chamber divided into two compartments by a diaphragm pierced by an aperture (see Fig. 4.5) of

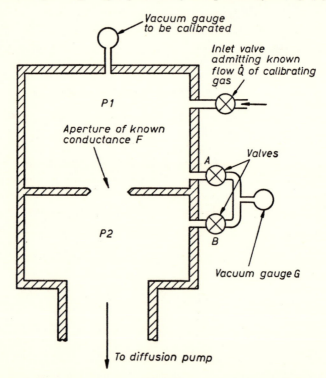

Fig. 4.5. Dynamic calibration of vacuum gauge

conductance F, calculated from its area,* the temperature and the molecular weight of the gas. The lower part of the chamber is connected to a diffusion pump of such speed that $P_1 \gg P_2$, P_1 and

* The correction to F, due to finite ratio of aperture area to diaphragm area is discussed in ref. 4.28. F itself is calculated from equation 2.7.

P_2 being the pressures above and below the aperture. The gauge to be calibrated is connected to the upper compartment, at pressure P_1.

In the steady state

$$\dot{Q} = F(P_1 - P_2)$$

or
$$P_1 = \frac{\dot{Q}}{F} \frac{1}{(1 - P_2/P_1)} \qquad (4.8)$$

P_2/P_1 is obtained from a gauge G which can be connected, by valves A and B, to either side of the aperture at will. Since $P_2/P_1 \ll 1$, it need not be determined with great accuracy, and G need not be calibrated in absolute terms since it only furnishes a ratio.

In the steady state, there are no adsorption errors, but it is necessary that outgassing should be small compared with $\dot{Q}$. For details ref. 4.29 may be consulted.

§ 4.10. Thermal transpiration and pressure measurement

Consider two chambers 1 and 2 at temperatures T_1 and T_2 separated by a small orifice at which the temperature discontinuity occurs.

If the pressures on the two sides are both so low that gas/gas collisions are negligible compared with gas/wall collisions, the molecules going from 1 to 2 will have a mean velocity $\bar{c}_1$, appropriate to T_1, and those from 2 to 1 will have $\bar{c}_2$ appropriate to T_2.

In the steady state the net flux through the orifice between the chambers must be zero.

Hence (see Appendix I, § 6),

$$n_1 \bar{c}_1 = n_2 \bar{c}_2 \qquad (4.9)$$

Thus, from equation (3) of Appendix I

$$\frac{n_1}{n_2} = \left(\frac{T_2}{T_1}\right)^{1/2} \qquad (4.10)$$

But (see equation (5) of Appendix I)

$$P = nkT$$

$$\therefore \frac{P_1}{P_2} = \frac{n_1 T_1}{n_2 T_2}$$

$$= \left(\frac{T_1}{T_2}\right)^{1/2} \qquad (4.11)$$

On the other hand, at high pressures where gas/gas collisions are predominant and continuous fluid mechanics apply, to a good approximation

$$P_1 = P_2 \qquad (4.12)$$

So, for the case of an aperture there is, as P increases, a continuous transition in behaviour between the regimes corresponding to equations (4.11) and (4.12). Note that equation (4.11) is valid for an *aperture* connecting two vessels at different temperatures, but not necessarily for a *tube*, which is the situation when a pressure gauge at room temperature is connected by tubulation to a work chamber at a different temperature. Edmonds and Hobson (ref. 4.30) working with He and Ne found equation (4.11) valid at low pressures for apertures, but not, in general, for pyrex tubes, for which it overestimates P_1/P_2 for $T_1/T_2 > 1$. They also give many references to earlier work and conclude that there is no rigorous theoretical or experimental basis for equation (4.11) as a general limit, except for apertures. The matter is reviewed in ref. 4.31. See also ref. 4.42.

§ 4.11. The determination of partial pressures

The pressure gauges described in previous sections give information on the total pressure of the gas phase, but none on its composition—which however is often useful (e.g. to identify the source of virtual or real leaks) and is crucial in some experiments, such as determination of low-pressure chemical kinetics.

4.11.1. *Partial pressures by mass spectrometry*

In nearly all types of mass spectrometer, the gas is ionized by means of an electron current from a hot filament, and the relative abundance of the resulting ions, separated according to the value of their mass/charge ratio, determined.

As the ionization probability is a function of the molecular species, the relative abundance of the ions is not the same as that of the neutral molecules from which they are derived. Also, the instruments do not present a spectrum of m alone. This can cause ambiguity, since for example singly ionized N and doubly ionized CO both give peak signals at $m/e = 14$.

In all types of mass spectrometer, the presentation of an ordered spectrum of m/e requires the motion of the ions to be controlled by the parameters of the instrument, undisturbed by chance collisions with gas molecules. Since the trajectory of the ions in a mass spectrometer is usually a few centimetres and their mean free path approximately equal to that of neutral molecules, there is an upper limit to the pressure of operation, usually of order 10^{-4} torr. (See Appendix I, § 9.)

4.11.2. *Important characteristics of mass spectrometers*

A mass spectrometer should have a large and linear response for a given ion, be able to scan the whole mass spectrum quickly, be bakeable (to avoid 'memory' of previous use), have little adsorption of gases to be analysed (which means minimum wall area), and little chemical reaction with the analysed gases, which demands minimum area and temperatures of the thermionic filament.

Ability to resolve neighbouring *lines* in the mass spectrum (called the mass resolution) is also of great importance.

The mass resolution R is defined in several different ways. A common definition is

$$R = \frac{M}{\Delta M} \qquad (4.13)$$

where M is the mass number* and $\frac{1}{2}\Delta M$ is the deviation on each

---

* M is the molecular weight of an ion divided by the number of electrons it has lost; e.g. M for doubly ionized CH_4 is 8.

side of M at which the intensity of the signal due to this mass number has fallen to f per cent of its peak value, where f is often either 0·1 or 1 per cent.

Suppose that an instrument is said to have $R = 30$, using the 1 per cent definition. This means, for example, that the signal due to $M = 45$ would fall to 1 per cent of its peak value at a distance of $\frac{3}{4}$ of a mass unit on either side of the peak value. Another often used definition of R is the mass number at which the height of the bottom of the valley between two peaks of equal height and one mass number apart is ten per cent of the height of either peak.

In assessing the resolution of an instrument it is important to know which definition of R is being used. For a comparison of the various definitions, see ref. 4.32.

4.11.3. *Types of mass spectrometer*

Mass spectrometers may be classified as static or dynamic. In the former type (in order to produce a peak signal for a particular value of m/e) the parameters of the instrument are held constant. In the dynamic instrument, one or more parameters are varied at a frequency which allows ions of particular value of m/e (unique to that frequency) to be collected.

The most popular mass spectrometer* of the static type is the magnetic deflexion instrument, in which ions produced by impinging electrons (as in a hot-cathode ionization gauge) are electrostatically extracted and caused to move in a circular arc of fixed radius R by a fixed magnetic field H (see Appendix II, § 16) at the termination of which is an ion collector. To each value of m/e there corresponds an extraction voltage V, and therefore initial velocity perpendicular to the fixed H, which will cause the ions to move in an arc of radius R, and thus reach the ion collector. The m/e spectrum is therefore scanned by varying V and observing the resultant ion current reaching the ion collector.

Another type of static instrument is the omegatron (see e.g. ref. 4.33), in which ions are produced by an electron beam parallel to a

---

* In cases where the presence of a magnetic field is compatible with other requirements of the situation.

constant magnetic field H under the influence of which they orbit in planes perpendicular to H with a frequency ω, which is determined only (see Appendix II, § 16) by their e/m. However, the orbital radius R is proportional to the square root of their kinetic energy. An r.f. electric field of variable frequency is applied in the plane of the orbits (i.e. perpendicular to H) and the ions can gain energy from this when the r.f. frequency becomes equal to ω. Thus, for each value of e/m there is a r.f. which will cause the path of the orbiting ions to spiral outwards until the ions impinge on a suitably placed collector. The current from this is amplified and monitored on an external recorder, so that the mass spectrum may be displayed as a plot of signal strength against radio frequency.

A type of dynamic instrument which does not require a magnetic field is the quadrupole mass spectrometer. In this, the ions are extracted from the ion source, and move down the longitudinal axis of an array of four cylindrical electrodes symmetrically placed with their centres $4r_0$ apart where r_0 is the radius of any one. Two diametrically opposite cylinders at a voltage U with respect to the ions source are connected together and the other two are also connected and have a common potential of $U + V \sin \omega t$. The frequency ω is a constant of the instrument, with the result that, for given values of U and V, only ions of a specific value of m/e can pass down the axis of the array and reach the collector. For other values of m/e they execute oscillations of increasing amplitude and are collected on the electrodes. Hence a scan of m/e is achieved by varying V and U, but the ratio V/U must be kept constant, since the resolution is determined by this. In a variant of this system, only two electrodes—a cylinder within a channel of rectangular cross-section—are used (ref. 4.34).

In the time-of-flight mass spectrometer, a pulse of ions is produced and accelerated by a controlled voltage, and therefore to a known velocity, and allowed to travel at this velocity down a tube of known length to an ion collector. If the ions are extracted from the ion source by a potential V and travel down a tube of length L at a velocity v to the collector,

$$eV = \tfrac{1}{2}mv^2 \qquad\qquad (4.14)$$

and t the time of flight is given by

$$t = L \left(\frac{m}{2eV}\right)^{1/2} \tag{4.15}$$

Hence ions of various m/e are formed into bunches and arrive at the collector after various time intervals, which can be displayed on a cathode ray oscillograph with a fast time sweep (ref. 4.35). One major advantage of this instrument is its very high rate of scan. Other types of mass spectrometer include the Favitron, and the linear accelerator type (see ref. 4.36).

For reviews of mass spectrometers, see Robinson (ref. 4.37), Leck (ref. 4.38), and Coleman (ref. 4.39). Coleman reviews the operating principles of five basic types and analyses in detail the specification and performance of magnetic deflexion instruments and quadrupole or monopole types.

4.11.4. *Measurement of partial pressures by sorption methods*

The desorption rate of a gas from a surface is dominated by the Boltzmann factor, $\exp(-q_d/kT)$, q_d and k being the desorption energy and gas constant per molecule (see Appendix II, § 19).

Suppose a surface located in a vessel containing a partial pressure of a gas is cooled to a temperature T such that $q \gg kT$. Adsorption of the gas occurs, and unless chemical combination occurs will be reversed on reheating. Because the temperature co-efficient of desorption is so large, owing to the Boltzmann factor, the desorption (which may be measured by a total pressure gauge in the system) becomes appreciable, rather suddenly, as the surface temperature rises, at a temperature T_d determined by q_d,[*] and hence by the particular gas desorbed. This is the basis of the simple but not very precise method of measuring partial pressures.

The temperature T_d is shown as a point of inflexion in the pressure/temperature curve, if the analysis is done in a closed system or, if the system is continually pumped, as a pressure *peak*,

---

[*] For any given sensitivity of pressure measurement there is a value of q_d/kT (and hence for a given q_d a value of T) at which the desorption becomes easily measurable.

the height of which, when the initial coverage was less than a mono layer, is a measure of the original partial pressure of the gas in the system. The value of T_d does not depend greatly on the rate of heating up of the surface, so that its value and the height of the peak allow both the adsorbed species and the partial pressure which give rise to it to be deduced.

The adsorption/desorption surface, whose temperature is caused to increase with time in a predetermined manner, is often a W filament. A typical procedure for determining partial pressures in a low-pressure vessel ($P \sim 10^{-10}$ torr) is first to 'flash' the W to about 2500°K to remove all adsorbed species. It is then allowed to cool to 300°K and, after a known period (t_c) at this temperature (to allow adsorption), is raised in a time of order 10^2 seconds to its original high temperature, the pressure and temperature being recorded. The amount of gas desorbed at T_d is a measure of the gas sorbed during t_c, while T_d itself determines the species. Prior calibration of the apparatus may be done by insertion of known amounts of known gases or with a mass spectrometer. For details ref. 4.40 should be consulted.

At relatively high pressures, say 10^{-2} torr, a crude analysis for H_2O and CO_2 is possible by cooling a Pyrex side tube to the vacuum vessel whose contents it is desired to analyse, and noting the temperature of the Pyrex surface at which rather sudden decreases of pressure occur. For example, H_2O is virtually completely condensed (i.e. removed from the gas phase) at temperatures between $-40°C$ and $-100°C$ and CO_2 between $-140°C$ and $-180°C$. In addition the amounts of any H_2 and O_2 present can be estimated by using heated Pd or Ag tubes to admit further quantities of H_2 or O_2 as necessary, promoting the $2 H_2 + O_2 = 2 H_2O$ reaction on a heated Pt filament, and determining the amount of water produced (ref. 4.41).

The above methods are laborious, limited in scope to those gases which are adsorbed at the lowest temperature available, and of low precision, but can be useful when convenience or expense preclude the use of a conventional mass spectrometer.

References

4.1. LECK, J. H., 1964, *Pressure Measurement in Vacuum Systems*, 2nd Edition, pp. 3 *et seq.* (Chapman & Hall Ltd).

4.2. ELLIOTT, K. W. T., WOODMAN, D. M., and DADSON, R. S., 1967, *Vacuum*, **17**, pp. 439–444.

4.3. ROTHE, E. W., 1964, *J. Vac. Sci. and Tech.*, **1**, pp. 66–68.

4.4. TUNNICLIFFE, R. J., and REES, J. A., 1967, *Vacuum*, **17**, pp. 457–459.

4.5. ROBERTS, J. K., and MILLER, A. R., 1955, *Heat and Thermodynamics*, 4th Edition, Chapter 11, Section 6 (Blackie and Son Ltd).

4.6. LECK, J. H., 1964, *Pressure Measurement in Vacuum Systems*, 2nd Edition, p. 58 (Chapman & Hall Ltd).

4.7. CLEAVER, J. S., 1967, *J. Sci. Inst.*, **44**, p. 969.

4.8. STECKELMACHER, W., 1965, *J. Sci. Inst.*, **42**, pp. 63–76.

4.9. ALLSOPP, H. J., BRADSHAW, F. J., STANFORD, R. H., and WADSWORTH, N. J., 1969, *Vacuum*, **19**, pp. 61–67.

4.10. DUSHMAN, S., Edited LAFFERTY, J. M., 1962, *Scientific Foundations of Vacuum Technique*, pp. 297 *et seq.* (John Wiley and Sons Inc.).

4.11. GREEN, M., and LEE, M. J., 1966, *J. Sci. Inst.*, **43**, pp. 948–949.

4.12. DUSHMAN, S., Edited LAFFERTY, J. M., 1962, *Scientific Foundations of Vacuum Technique*, pp. 301 *et seq.* (John Wiley and Sons Inc.).

4.13. LECK, J. H., 1964, *Pressure Measurement in Vacuum Systems*, 2nd Edition, p. 91 (Chapman & Hall Ltd).

4.14. LECK, J. H., 1964, *Pressure Measurement in Vacuum Systems*, 2nd Edition, p. 103 (Chapman & Hall Ltd).

4.15. LEWIN, G., 1965, *Fundamentals of Vacuum Science and Technology*, p. 96 (McGraw-Hill Book Co.).

4.16. COBIC, B., CARTER, G., and LECK, J. E., 1961, *Vacuum*, **11**, pp. 247–257.

4.17. TURNBULL, A. R., BARTON, R. S., and RIVIÈRE, J. C., 1962, *An Introduction to Vacuum Technique*, p. 89 (George Newnes Ltd).

4.18. HELMER, J. C., and HAYWARD, W. H., 1966, *Rev. Sci. Inst.*, **37**, pp. 1652–1654.

4.19. FITCH, R. K., and THATCHER, W. J., 1968, *J. Sci. Inst.*, March. pp. 317–319.

4.20. MOURAD, W. G., PAULY, T., and HERB, R. G., 1964, *Rev. Sci. Inst.*, **35**, pp. 661–665.
4.21. LECK, J. H., 1964, *Pressure Measurement in Vacuum Systems*, 2nd Edition, p. 115 (Chapman & Hall Ltd).
4.22. TURNBULL, A. R., BARTON, R. S., and RIVIÈRE, J. C., 1962, *An Introduction to Vacuum Technique*, p. 72 (George Newnes Ltd).
4.23. STECKELMACHER, W., 1951, *Vacuum*, **1**, pp. 266–282.
4.24. MAIR, W. N., 1967, *Royal Aircraft Establishment Tech. Rep.*, 67024.
4.25. LECK, J. H., 1964, *Pressure Measurement in Vacuum Systems*, 2nd Edition, pp. 35 *et seq.* (Chapman & Hall Ltd).
4.26. DUSHMAN, S., Edited LAFFERTY, J. M., 1962, *Scientific Foundations of Vacuum Technique*, pp. 224 *et seq.* (John Wiley and Sons Inc.).
4.27. MEINKE, C., and REICH, G., 1967, *J. Vac. Sci. and Tech.*, **4**, pp. 356–363.
4.28. BUREAU, A. J., LASLETT, L. J., and KELLER, J. M., 1952, *Rev. Sci. Inst.*, **23**, p. 683.
4.29. LECK, J. H., 1964, *Pressure Measurement in Vacuum Systems*, 2nd Edition, pp. 168 *et seq.* (Chapman & Hall Ltd).
4.30. EDMONDS, T., and HOBSON, J. P., 1965, *J. Vac. Sci. and Tech.*, **2**, pp. 182–197.
4.31. REDHEAD, P. A., HOBSON, J. P., and KORNELSEN, E. V., 1968, *The Physical Basis of Ultra-High Vacuum*, pp. 281 *et seq.* (Chapman & Hall Ltd).
4.32. DENNIS, N. T. M., and HEPPEL, T. A., 1968, *Vacuum Design*, pp. 97 *et seq.* (Chapman & Hall Ltd).
4.33. ROBINSON, N. W., 1968, *The Physical Principles of Ultra-High Vacuum Systems and Equipment*, pp. 111 *et seq.* (Chapman & Hall Ltd).
4.34. ROBINSON, N. W., 1968, *The Physical Principles of Ultra-High Vacuum Systems and Equipment*, p. 126 (Chapman & Hall Ltd).
4.35. ROBINSON, N. W., 1968, *The Physical Principles of Ultra-High Vacuum Systems and Equipment*, pp. 128 *et seq.* (Chapman & Hall Ltd).
4.36. ROBINSON, N. W., 1968, *The Physical Principles of Ultra-High Vacuum Systems and Equipment*, pp. 127 and 128 (Chapman & Hall Ltd).
4.37. ROBINSON, N. W., 1968, *The Physical Principles of Ultra-High Vacuum Systems and Equipment*, Chapter 4 (Chapman & Hall Ltd).

4.38. LECK, J. H., 1964, *Pressure Measurement in Vacuum Systems*, 2nd Edition, Chapter 7 (Chapman & Hall Ltd).

4.39. COLEMAN, J. W., 1967, *Technical Memorandum No. 110152 of the B.P. Research Centre*, Sunbury-on-Thames, Middlesex, England.

4.40. REDHEAD, P. A., HOBSON, J. P., and KORNELSEN, E. V., 1968, *The Physical Basis of Ultra-High Vacuum*, pp. 355 *et seq.* (Chapman & Hall Ltd).

4.41. CARPENTER, L. G., and MAIR, W. N., 1953, *Royal Aircraft Establishment Tech. Note*, No. Met. 174.

4.42. HOBSON, J. P., 1969, Surface smoothness in thermal transpiration at very low pressures. *Jour. Vac. Sci. and Tech.*, 6, pp. 257-259. (The thermal transpiration ratio for a *leached* Pyrex tube joining the hot and cold regions agrees closely with the ratio obtained with an aperture; in particular $P_1/P_2 = (T_1/T_2)^{\frac{1}{2}}$ is found for very low pressures.)

5

MATERIALS
AND THE
DESIGN OF APPARATUS

§ 5.1. Material properties and design requirements

Space does not permit detailed description of the properties of the many materials used in vacuum technology, for which reference may be made to existing authoritative sources, see, for example, Rosebury (ref. 5.1) and Espe (ref. 5.2). This chapter will indicate some considerations governing material choice, briefly mention a few common materials and vacuum accessories (such as valves), and outline a typical design procedure. Some notes on the troublesome and time wasting matter of leak detection are included.

The choice of materials of which a system is constructed is strongly influenced by its intended use. For example, if the system is an industrial one, and ruggedness is desired, metal is usually preferable to glass. In research laboratories, glass may be suitable, but the choice may be influenced by the relative availability of glass- and metal-working facilities. In a metal system, ease of leak-tight welding or brazing must be considered. Operating temperature is important; if part of the apparatus is at low temperature, elastomers which become brittle cannot be used—or stressed parts of metal liable to brittle fracture. A low thermal conductivity may be desirable to limit heat conduction from the warm to the cold parts of the apparatus (and hence to conserve refrigerant) or a high one to improve isothermality. At high temperatures, vapour pressure, which is a strong function of temperature (see Appendix

6

II, § 3), precludes the use of some metals (e.g. zinc distils out of brass) and thermal degradation the use of organic substances. Outgassing rate (see § 5.8) is always an important property in high vacuum apparatus, since it limits both the final pressure attainable and the time taken to reach it.

§ 5.2. Glass and glass-to-metal seals

Borosilicate (so-called 'hard') glasses are generally used in preference to 'soft' ones, because they have a low coefficient of linear expansion (usually between 3 and 6 × 10^{-6} °C^{-1}) and therefore do not easily crack when subjected to a thermal gradient, and a relatively high (up to about 450°C) service temperature. Soft glasses are often used in large-scale production—e.g. soda lime glasses as the envelopes for small lamps and thermionic valves.

The most common hard glass used in vacuum work is Pyrex (Corning 7740) and many vacuum gauges are now made in glasses which seal directly to it. Otherwise, a graded seal (consisting of several glasses of expansion coefficient decreasing gradually between that of the gauge and the Pyrex of the vacuum apparatus to which it is connected) is used. Pyrex apparatus can be cooled with L.N$_2$ (77°K) without much risk of cracking and has an upper service temperature of about 450°C, but Bills and Evett (ref. 5.3) warn that it may be a disturbing factor in certain *clean-surface* scientific experiments, as it appears to decompose at temperatures above about 350°C, liberating H_2O and (more serious) certain contaminants not readily pumped out of a vacuum system. Donaldson also (ref. 5.4) calls attention to certain undesirable effects. However, for work which does not call for ultra-cleanness, Pyrex remains a most useful material.

Fused silica (i.e. amorphous glass derived from crystalline quartz) has a service temperature up to about 1000°C, because of its low coefficient of linear expansion ($\sim$0·5 × 10^{-6} °C^{-1}) is virtually unbreakable by thermal gradient, and is a much better electrical insulator than Pyrex. It can be sealed to the latter via a number of intermediate glasses of carefully graded expansion coefficients.

The uses of glass/metal seals in vacuum systems include the following:

(*a*) To make a junction in the vacuum envelope between the two materials.

(*b*) To introduce electrical leads into glass vacuum envelopes.

(*c*) To introduce electrical leads into metal vacuum envelopes, the glass serving as an insulator.

Purpose (*c*) can also be served by insulators other than glass. See e.g. ref. 5.5.

In making glass/metal seals it is necessary that the glass should bond to the metal or its oxide, and either that their expansion coefficients should not be greatly different or that the metal should, by elastic or plastic deformation, conform to the dimensions of the glass as its temperature changes. An example of approximately matched expansion is the W/Pyrex seal, often used for introducing electrical leads into vacuum systems. Another example of this is the Kovar/Kodial seal—the former being an alloy of Ni, Co, and Fe which matches Kodial glass, which itself can be sealed to Pyrex via one intermediary glass—the *graded seal* technique.

The alternative *metal-conforming* method is exemplified by the Housekeeper seal, in which a feather edge of a copper tube is sealed into a Pyrex tube. A similar technique has been used for joining Pyrex and stainless steel. Making metal/glass seals is a highly skilled technical job. Fortunately, most types are commercially available. For an excellent review of possibilities, see the book by Green on constructing small vacuum systems (ref. 5.6). Holland's book on the properties of glass surfaces (ref. 5.7) contains many data useful in vacuum technology.

§ 5.3. Metals

The most commonly used metal for vacuum envelopes are brass, copper, and stainless steel. Brass, being an alloy of Cu and Zn, cannot be used above 100 to 150°C because at high temperatures *in*

vacuo the Zn distils out.* Copper, having a much lower vapour
pressure, can be used to much higher temperatures *in vacuo*, but if
used as part of a vacuum envelope, its outer surface, exposed to air,
will be seriously oxidized above a few hundred degrees centigrade.
It is particularly useful when heat has to be conducted to or from
certain parts of the apparatus. Oxygen-free high-conductivity
copper (O.F.H.C.) should generally be used, since copper containing oxygen reacts at temperature with H_2 to produce H_2O, and
some grades of ordinary copper contain impurities which very
seriously reduce its thermal conductivity, e.g. 0·4 per cent of
arsenic, often deliberately added to improve mechanical properties,
may reduce both electrical and thermal conductivities to less than
half those of pure copper.

A suitable steel for welded vacuum vessels is B.S. 970 EN 58G
(American equivalent AISI 347). This is a non-magnetic S.S. of
the 18/8/1 type, the latter component being Nb, whose purpose is
to combine with any C present and prevent formation of chromium
carbide, which by depleting the Cr content of the grain boundaries
leads to weld decay. Where conditions are not so stringent as
regards freedom from small leaks and minimum magnetic per-
meability is not a requirement, B.S. 970 EN 58B (American equivalent AISI 321) (in which Ti is substituted for Nb and the Ni
content is lower) may be used.

Mild steel is sometimes used for vacuum envelopes, but because
it rusts when let down to atmosphere, thereby increasing its outgassing rate in subsequent service, it is usually nickel- or chromium-
plated. Also at low temperatures ordinary mild steel should be used
with caution in vacuum vessels owing to risk of brittle fracture
caused by 'locked up' stresses.

Turnbull, Barton, and Rivière (ref. 5.8) give a useful list of the
physical properties of metals of interest in vacuum work.

§ 5.4. Plastics

Plastics, in the sense of rigid organic materials (as opposed to
elastomers dealt with in § 5.6), are little used as materials of con-

---

* Similarly, high vapour pressure precludes the use of hard solders
containing Cd.

struction in vacuum technology. However Nimrod—the 7 G.E.V. proton synchrotron at the Rutherford Laboratory, Chilton, Berkshire, England, more fully described in § 7.2—deserves special mention. The double-walled vacuum chamber is made from glass fibre epoxy laminate, chosen because of absence of eddy current loss in intense and varying magnetic fields, resistance to a high level of radiation before seriously deteriorating, and low outgassing rate.

Minor uses of plastics include PVC tubing for vacuum pipe lines at pressures greater than 10^{-4} torr (ref. 5.9), and synthetic resin adhesives as vacuum-tight cements in experimental vacuum apparatus (e.g. see ref. 5.10). Barton and Govier (ref. 5.11) have studied the outgassing of certain plastics, including Araldites, Mycalex, Perspex, and P.T.F.E., analysing the evolved species with a mass spectrometer. For useful information on epoxy resins (and on AgCl used as a high temperature cement) see Green (ref. 5.12). Kendall and Zabielski (ref. 5.13) have published valuable data on high-temperature ($\sim 300°C$) insulating adhesives for vacuum applications.

§ 5.5. Waxes and greases

Waxes are useful for making vacuum seals in semi-permanent experimental apparatus. They should have a low vapour pressure and a softening point above the highest likely service temperature, and should not decompose when heated during application. The Shell Chemical Co. Ltd market a range of vacuum waxes with melting points from 45° to 85°C and vapour pressures, *after evolution of dissolved air*, of 10^{-3} torr at 180°C.

Greases are used as sealants and lubricants in glass vacuum taps and ground joints. Again a number of greases are available with vapour pressures at 20°C ranging from 10^{-3} to 10^{-8} torr after the evolution of dissolved air. This qualification is important, since the greases in a newly assembled vacuum system are often sources of considerable gassing—a fact demonstrated by the rise in reading of a pressure gauge when a newly greased tap is turned.

According to Allsop (ref. 5.14) vapours from vacuum waxes and greases may be 'cracked' on a hot Pt filament, producing significant quantities of low-molecular-weight gases.

The elastomer O-rings of demountable couplings are often lightly greased to improve leak tightness. After so doing, care should be taken not to allow the greased O-rings to pick up dust particles or fibres which may spoil their seating and degrade rather than improve their vacuum integrity.

§ 5.6. Demountable couplings and motion in vacuo

Demountable couplings are usually circular in cross-section, and their vacuum tightness depends upon the deformation of an elastomer or metal gasket (in the latter irreversible) compressed by some mechanical device. The gaskets are either elastomers, used in systems not subjected to severe baking, or metals, used in U.H.V. systems baked to temperatures up to about 450°C.

In designing an apparatus or setting up a vacuum laboratory it is well to study carefully the various types of coupling commercially available, and having chosen one, to standardize on that. The unnecessary use of several types can lead to much expense and loss of time.

The subject of couplings, though mundane, is very important in practice. An attempt to describe the various options would be out of place in a small book, and superficial treatment useless. But a few broad points with references to sources of detailed information are given below.

The important qualities in elastomers are maximum service temperature, impermeability to gases, outgassing rates, resistance to oil, and mechanical properties, especially degree of resistance to permanent 'set' under compression.

A common elastomer used in vacuum seals is nitrile rubber (i.e. a copolymer of butadiene and acrylonitrile) with a maximum service temperature of about 80°C. For temperatures up to 200°C, a fluorinated elastomer marketed as Viton A can be used, but as this tends to acquire a permanent set it should preferably be confined in a groove or recess of cross-section area not more than 5 per cent greater than that of the gasket itself. For an excellent account of elastomer seals and associated couplings based on extensive firsthand experience, see Chapter 5 of ref. 5.15.

O-ring gaskets (i.e. gaskets with a circular cross-section) usually have a leak rate less than 10^{-2} lusec, in e.g. pipe couplings of a few cm diameter. There are numerous elastomer type couplings. For example the Vac-Lok system includes an extensive and well-designed range of couplings of various diameters, with associated valves, elbow bends and T-junctions, and adaptors for changing from one diameter of tube to another.

Metal demountable seals depend upon the plastic irreversible flow of metal gaskets, compressed by forces exerted on metal flanges by bolts. The metals should have low vapour pressure, high melting point, and reasonable resistance to oxidation.

In the Varian Conflat seal, an originally flat copper ring gasket is compressed between projections in two similar stainless steel flanges, the amount of plastic flow, and therefore stress relief in the copper, being limited by the geometry of the flanges, which are bolted up until they come in contact.

In another type, a copper gasket of diamond cross-section is compressed between two stainless steel flanges, in one of which there is a circular groove, whose depth is 85 per cent of the apex-to-apex dimension of the gasket. The mating surfaces of the flange make contact when bolted up. In a simpler but reputedly less satisfactory variant of this type, the faces of the two flanges are identical, both being plane.

A simple and popular seal consists of a gold wire gasket compressed between bolted flanges to about half its original diameter. In all the above seals, the parts of the flanges in contact with the deformable gasket should have a fine machine finish and, in particular, should be free from radial scratches.

The seals mentioned above are merely a selection from a considerable number of types, about the relative merits of which there is as yet no universal agreement. A worker contemplating the use of bakeable metal seals should consult the work of authorities such as

Robinson (ref. 5.16), Turnbull, Barton, and Rivière (ref. 5.17), and Lewin (ref. 5.18).

In vacuum systems not subject to baking, transmission of rotational motion from outside may be by means of greased rotatable cone-and-socket joints or by a highly polished metal shaft passing through a greased and suitably supported elastomer.

For baked systems, a circular pendulous motion of a shaft may be transmitted by a metal bellows, usually of stainless steel. By moving the outer (atmospheric) end, the inner (vacuum) end of the shaft can be manipulated so as to cause the rotation of a bearing itself wholly within the vacuum. For details of this and other like devices, see refs. 5.19 and 5.20.

§ 5.7. Vacuum taps and valves

For small laboratory apparatus, greased glass stopcocks are still much used. They consist of a conical glass key carefully ground into an outer glass barrel, to which are fused tubular glass side arms. In one form the key is traversed by a diametral bore, the tap being open when the bore is in line with side arms. In use continued rotation of the tap tends to produce circular streaks in the grease, which may eventually lead to partial short circuiting of the tap, even when it is nominally closed, with the axis of the bore perpendicular to side arms. This particular type of leak is obviated in a second form in which the bore passes obliquely across the key, the two side arms being offset.

Glass stopcocks with bores of more than about 2·5 cm are unwieldy and often become very difficult to turn as the grease ages. By means of glass/metal tubular seals (see § 5.2), it is possible to incorporate glass stopcocks in otherwise all-metal systems, and this is sometimes advantageous, since glass stopcocks greased with a low-vapour-pressure grease probably outgas less than metal valves of the diaphragm type.

In the latter, the seal between the two sides of the valve is made by an elastomer diaphragm moved into the closure position by an externally operated screw mechanism. In a variant of this, the closure is effected by the movement perpendicular to its own plane of a circular plate carrying an O-ring in a groove on its perimeter, isolation from the external atmosphere being by means of a bellows.

Other types of non-bakeable valve include the gate type, in which closure between the two sides is made by the movement of a horizontal plate in its own plane, the seal being an elastomer, and the quarter swing (butterfly) type, in which a disc is caused to rotate about a horizontal shaft crossing the diameter of the circular aperture to be closed.

Both gate and butterfly types are often used immediately above the mouth of a diffusion pump or above the water-cooled baffles

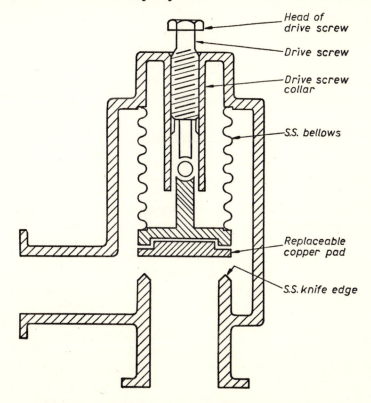

Fig. 5.1. Bakeable all-metal valve [schematic]

(see § 3.3.1) if any are fitted. A third type, commonly used for the same purpose is the plate valve, in which a horizontal circular plate of metal is located immediately above the mouth of the diffusion pump and can be moved vertically by means of a lever mechanism operated from outside. In the closed position, the plate rests on the circular mouth of the diffusion pump, the vacuum seal being provided by the peripheral O-ring; in the open position, the plate is suspended a centimetre or so above the pump mouth, thus itself forming a baffle which tends to prevent oil vapour from the diffusion pump from going unimpeded to the work chamber, but allows pumping to proceed without undue reduction of speed.

Valves in apparatus to be outgassed by baking at above 300°C must be of all-metal construction. Fig. 5.1 shows the principle of one common type, in which vacuum-tight closure is obtained by forcing a pad of a soft metal, such as copper, on to the mouth of a sharp-edged orifice of a hard metal, such as stainless steel. The driving mechanism, usually a threaded shaft, is at atmospheric pressure and is separated from the evacuated space by a metal bellows.

§ 5.8. Outgassing

When a solid is first exposed to a vacuum, gas is desorbed, and the desorption continues indefinitely, but at a decreasing rate. The released gas may in principle be considered as consisting of two kinds, that desorbed from the surface and that which diffuses from the interior, but since the surface gas is constantly replenished by that coming from the interior, the distinction is somewhat academic. However, it is often true that the initial surface-desorbed gas is mainly H_2O, at any rate for metals and glasses. There exists a considerable body of data on the rate of outgassing of various materials as a function of time. The rate, usually expressed in torr litres sec^{-1} cm^{-2}, decreases in a roughly exponential manner with time. For example, a typical elastomer O-ring might have an initial rate of 10^{-6} (in the above units) decreasing by a factor of 2 at the end of 5 hours and another factor of 2 at the end of 10 hours. Corresponding figures for slightly rusty mild steel might be 10^{-6} when first exposed to vacuum, with a decrease by a factor of about 10 in the first 5 hours and by a similar factor in the next 5 hours. Stainless steel of the 18:8:1 type might have an initial rate about a decade lower, but the rate decays in about the same way as for mild steel.

Outgassing/time curves for various constructional materials are given by Power (ref. 5.21) and by Dennis and Heppell (ref. 5.22), who themselves give references to further data.

The outgassing rate of a given material is strongly dependent on its previous surface treatment. For instance, the rate after 1 hour for 'raw' stainless steel is diminished by a factor of 5 by electro-polishing (Dennis and Heppel, *loc cit.*). Also, outgassing increases rapidly with temperature, so that even a mild bake at 100°C greatly reduces the rate when the material returns to room temperature. Hence the vacuum technologist's adage: 'A little temperature is worth a lot of time.'

§ 5.9. Design procedure

In a design it is necessary to consider:

Physical size of work chamber
Temperature of work chamber
Electrical connection to inside of work chamber
Mechanical motions to inside of work chamber
Required precision of pressure measurement
Degree of vacuum required
Pump-down time to required vacuum
Skill of the intended operator

If a work chamber of the bell-jar type (i.e. a bell jar mounted on a flat base plate, an L-section circular elastomer being interposed as a seal) is contemplated, the maximum diameter should not exceed 30 cm if the bell jar is glass*; above this diameter the bell jar will be metal—usually stainless steel. If the power input to the contents is more than a few tens of watts, the base plate (and possibly the bell jar) is usually water-cooled.

If the vacuum envelope is of the demountable gasket type, e.g.

---

* Glass bell jars should have external screens to limit danger following 'implosion'.

bell jar or 'top hat' on a base plate, the electrical and mechanical connections are conveniently brought in through a 'service' collar, which is interposed between the base plate and the main vacuum envelope. The advantage of this arrangement is that, if a modification to electrical or mechanical inputs is required, it can be effected by modifying the collar without interfering with the base plate, which is usually connected to the pump via a valve and (often) by a L.N$_2$-cooled vapour trap.

For pressure measurement, a Pirani gauge is useful to cover the range before it is safe (i.e. 10^{-3} torr or preferably less) to switch on an ionization gauge. If only a rough indication in the range 10^{-2} to 10^{-6} torr is required, the Penning gauge, being cold-cathode and therefore immune from burn out, has much to commend it.

If the apparatus is to be operated by unskilled personnel, safety devices to guard against accidents, such as the switching on of diffusion pumps without backing pumps or omitting to turn on cooling water, are desirable. If the required pressure in the work chamber is comparatively high, say not much less than 10^{-1} torr, the pump-down time of the empty chamber may be calculated by the procedure of § 2.8. If, however, pressures in the range 10^{-3} to 10^{-6} are required, virtual leak (i.e. desorption) becomes important and a different method must be used.

The following numerical example is given merely to illustrate *method*, using figures which may give some 'feel' for orders of magnitude.

Suppose that the problem is to choose pumping equipment for a small 'top hat' type stainless steel vertical cylindrical chamber of height 60 cm and diameter 30 cm. The service collar carrying electrical leads and other services is sandwiched between two elastomer gaskets. Above the collar is the bottom flange of the top hat, and below it the base plate of the apparatus, which latter is connected to a diffusion pump via a baffle valve. Assume, as is likely, that a combination of oil diffusion pump and mechanical backing pump has been chosen. It is desired to attain a vacuum of about 10^{-5} torr in 1 hour after exposure of the chamber to atmospheric air. What effective pumping speed at the *entry port* in the base plate must be provided?

The following calculation is valid only for a clean and empty chamber. Allowance for the outgassing of any work load must be made subsequently. Even so, two assumptions must be made regarding (i) the outgassing rate of the S.S. inner surface after 1 hour's pumping (virtual leak) and (ii) the real inleak of air from outside, due perhaps to imperfections in the elastomer seals and to leaks in the service entries in the service collar. For (i) we shall take 2×10^{-8} torr litre cm^{-2} sec^{-1} after 1 hour (ref. 5.23) and for (ii) 10^{-2} lusec, i.e. 10^{-5} torr litre sec^{-1}. Both these assumptions are guesses. As to (i) the outgassing of S.S. depends greatly on its history and surface finish and is *not* a 'constant' of the material, while (ii) is a figure which experience shows to be a fairly large leak, and perhaps, therefore, a safe upper limit for reasonably well-constructed apparatus.

The volume of the top hat is about 40 litres and its internal surface about 0·7 metres^2. The virtual leak after 1 hour is therefore 14×10^{-5} torr litre sec^{-1} and the real leak 1×10^{-5} torr litre sec^{-1}, so the assumed value of the real leak will only increase P_1 (the pressure after 1 hour's pumping) by about 7 per cent above the no-real-leak value.

Hence, if S_c is the required pump speed at the work chamber

$$P_1 S_c = \text{throughput in } PV \text{ units (equation 5.2) per unit time}$$
$$= 15 \times 10^{-5} \text{ torr litre sec}^{-1}$$

and $\quad S_c = 15$ litres sec^{-1}

Also P_∞, the pressure after infinite time, when the virtual leak had become zero would be given by $P_\infty \times 15 = 1 \times 10^{-5}$, or $P_\infty = 6·7 \times 10^{-7}$ torr. Choice of backing pump depends on several factors. In any case, it must retain pumping capability down to the maximum backing pressure at which the diffusion pump will operate. Assuming that oil diffusion pumps retain a reasonable fraction of their maximum throughput (see § 3.3.4) at an inlet pressure of order 10^{-2} torr, and require a backing pressure of $\sim 3 \times 10^{-1}$ torr, a rough guide to the speed S_b of the backing pump required to match conditions of maximum throughput during pump-down would be

$$S_c \times 10^{-2} = S_b \times 3 \times 10^{-1}$$

or
$$\frac{S_b}{S_c} = \frac{1}{30}$$
(5.2)

A considerably smaller ratio of S_b to S_c than that indicated by equation (5.2) is often quite satisfactory if maximum throughput during the early stages of pump-down is not necessary. The intention here is merely to indicate the type of reasoning, the actual figures used depending on the pump-down schedule required. Applying equation (5.2) to the present case gives $S_b = \frac{1}{2}$ litre sec^{-1} or 30 litres min^{-1}, the latter unit being conventional for backing pumps. The pump-down time from 760 torr to 10^{-1} torr (down to which pressure outgassing may be assumed to be insignificant) would be correctly predicted by equation (2.16) as about 12 minutes,* assuming S_b constant, and not significantly reduced by impedance of backing line.

In contrast to this, equation (2.16) incorrectly gives a time to pump-down from 10^{-2} torr (the input pressure at which the diffusion pump becomes effective) to 10^{-5} torr of about 18 seconds, as against the real value of 1 hr, and so is optimistic by a factor of nearly 200—because it neglects outgassing.

The required speed, given by equation (5.1) is the speed at the entry port of the work chamber, and if a baffle is interposed between work chamber and diffusion pump the effective speed is usually reduced by about a factor of 2. Allowing another factor of 2 for contingencies, a reasonable estimate of unbaffled diffusion pump speed for the case of the clean empty chamber would be 60 litres sec^{-1}, plus (as explained above) an allowance for degassing of chamber contents. It thus appears that precise calculation of required pump capacity is in general not possible unless experience of the degassing characteristics of the chamber contents is available. However, the type of procedure outlined at least enables an estimate of the *minimum* reasonable pumping capacity, to which

* This would probably be satisfactory in this case, as the total time is of order 1 hour. But if the time consumed in the low pressure part of the pump-down were reduced (e.g. by not requiring so low a final vacuum), it might be worth while to reduce the 'roughing out' time by use of a larger backing pump. Thus, especially where quick cycling is required, economic considerations of capital cost versus rate of output influence design.

must be added an allowance for the outgassing of the contents of the work chamber. In the absence of first-hand data on this, a preliminary guess may be made with the aid of ref. 5.24, which itself refers to other sources of data. For instructive examples of pump-down calculations see Steinherz (ref. 5.32)

§ 5.10. Leak detection

There are methods for the detection of a leak but not of its location, methods which indicate the location using equipment forming part of the plant, and methods using instruments applied specially to the plant for the detection and location of its leaks. The subject has been reviewed by Turnbull (ref. 5.30).

5.10.1. *Detection of presence, but not location*

One procedure consists simply in determining the pressure/time curve of the suspected apparatus when isolated from its pumps. If the pressure rises linearly with time and appears to continue to do so indefinitely, the leak is probably real rather than virtual. If L.N$_2$ is applied to a cold finger on the apparatus and most of the gas accumulated (say) overnight disappears, then at any rate most of the leak is virtual and probably consists of desorbed H$_2$O or other condensible gas not forming a major constituent of the external atmosphere, but derived from walls and contents of the work chamber.

Another procedure (ref. 5.25) depends on the fact that the speed of a rotary pump is much greater for air than for a condensible vapour.

5.10.2. *The bubble method*

The apparatus to be tested is immersed in or painted with water, to which soap or some other surface-tension-reducing agent has been added, and internally pressurized by air. A leak, if sufficiently large, causes a bubble to form at its outlet. A brush should be used to clear the submerged surface of bubbles formed on immersion, which would confuse the issue. According to ref. 5.26, the minimum detectable leak is 10^{-1} lusec.

5.10.3. *Use of Tesla discharge*

The discharge of a Tesla coil (see § 4.6.2) can locate large leaks (order of 1 lusec) in glass systems if its high tension probe is passed slowly over suspected parts. The normally bushy discharge jumps to the pinhole leak through which it enters the vacuum system, and in doing so the leak is shown up as a brilliant white spot or line in the glass.

The Tesla discharge can also be used to detect a leak by holding it near the glass part of the apparatus causing a luminous glow within, while the suspected leaking areas (which may be glass or metal) are painted with a volatile hydrocarbon such as ether. When the leak is reached, the hydrocarbon enters the apparatus and the colour changes from the purple-red characteristics of air to a greyish blue. The virtual leak from the H_2O-laden walls of a newly evacuated apparatus will often give a blue-white discharge which may make difficult the interpretation of a painting-with-hydro-carbon search for a real leak.

5.10.4. *Use of instrumentation already part of the apparatus*

If the leak or leaks are such as to prevent the attainment of pressures lower than about 10^{-3} torr, the pressure is in the range of a Pirani gauge. As the response of this gauge depends upon the nature of the gas as well as its pressure, leaks may be located by probing the outside of the apparatus with a fine jet of a gas to which the gauge response differs from its response to air, and watching the gauge for a sudden change of reading.

If the ratio between leak rate and pump speed is small enough to cause the equilibrium pressure to be $\not> 10^{-4}$ torr, the hot-cathode ionization gauge can, without undue risk of filament burn out, be used to locate leaks in the same way as the Pirani gauge. If He is the search gas, the indication of the gauge drops when the leak is located; but if butane is used, it rises.

If evacuation is via a sputter ion pump, the pump itself may be used as a leak detector, since the ion pump current depends not only on the pressure of the gas in the pump but also on its species.

He gives the largest increase in current (since it is less well pumped than air) and O_2 the largest decrease. For detailed analysis of this method, see ref. 5.27.

5.10.5. *Use of special instrumentation*

There are a number of special instruments for leak detection, of which two types may be mentioned here:

(*a*) Types in which there is, between a special gauge and the rest of the vacuum system, a barrier impermeable to air and to all gases likely to be found in the system, but permeable to the probe gas applied to the apparatus from outside. The gauge then indicates when the probe gas enters the system. Examples of this type are the palladium barrier detector (ref. 5.28) or a cooled charcoal barrier (ref. 5.29), both of which are permeable to H_2 but not to most other gases.

(*b*) The mass-spectrometer type, in which the probe gas is usually He and the spectrometer (tuned for He) is connected to the apparatus being tested for leaks, a signal being obtained when the helium impinges on the entrance to the leak. Sensitivity of detection is high (figures of order 10^{-8} lusec are quoted), but comparison of the sensitivities of various instruments must be made cautiously unless the exact conditions of use are specified. For instance, the time of application of He to produce a given signal is clearly relevant (ref. 5.30). For a critical discussion of sensitivity of leak detection, see ref. 5.31.

References

5.1.   ROSEBURY, F., 1965, *Handbook of Electron Tube and Vacuum Technique* (Addison-Wesley Publishing Inc.).

5.2.   ESPE, W., 1966–8, *Materials of High Vacuum Technology*, Vols 1–3 (Pergamon Press Ltd).

5.3.   BILLS, B. G., and EVETT, A. A., 1959, *Jour. App. Phys.*, **30**, pp. 564–567.

5.4.   DONALDSON, E. E., 1962, *Vacuum*, **12**, pp. 11–14.

5.5.   KOHL, W. H., 1964, *Vacuum*, **14**, pp. 333–354.

7

5.6. GREEN, G. W., 1968, *The Design and Construction of Small Vacuum Systems*, Chapter 3 (Chapman & Hall Ltd).

5.7. HOLLAND, L., 1961, *The Properties of Glass Surfaces* (Chapman & Hall Ltd).

5.8. TURNBULL, A. H., BARTON, R. S., and RIVIÈRE, J. C., 1962, *An Introduction to Vacuum Technique*, Chapter 5 (George Newnes Ltd).

5.9. HARTMAN, R. L., 1967, *Rev. Sci. Inst.*, **38**, p. 831.

5.10. SAYERS, J. F., 1960, *Jour. Sci. Inst.*, **37**, pp. 203–205.

5.11. BARTON, R. S., and GOVIER, R. P., 1965, *Jour. Vac. Sci. and Tech.*, **2**, p. 113–122.

5.12. GREEN, G. W., 1968, *The Design and Construction of Small Vacuum Systems*, pp. 77–78 (Chapman & Hall Ltd).

5.13. KENDALL, B. R. F., and ZABIELSKI, M. F., 1966, *Jour. Vac. Sci. and Tech.*, **3**, pp. 114–119.

5.14. ALLSOPP, H. J., 1961, *Vacuum*, **11**, p. 39.

5.15. TURNBULL, A. H., BARTON, R. S., and RIVIÈRE, J. C., 1962, *An Introduction to Vacuum Technique* (George Newnes Ltd).

5.16. ROBINSON, N. W., 1968, *The Physical Principles of Ultra-High Vacuum*, Chapter 3 (Chapman & Hall Ltd.).

5.17. TURNBULL, A. H., BARTON, R. S., and RIVIÈRE, J. C., 1962, *An Introduction to Vacuum Technique*, Chapter 5 (George Newnes Ltd).

5.18. LEWIN, G., 1965, *Fundaments of Vacuum Science and Technology*, Chapter 7 (McGraw-Hill Book Co.).

5.19. TURNBULL, A. H., BARTON, R. S. and RIVIÈRE, J. C., 1962, *An Introduction to Vacuum Technique*, pp. 127 *et seq.* (George Newnes Ltd).

5.20. GREEN, G. W., 1968, *The Design and Construction of Small Vacuum Systems*, pp. 128 et seq. (Chapman & Hall Ltd).

5.21. POWER, B. D., 1966, *High Vacuum Pumping Equipment*, pp. 392 *et seq.* (Chapman & Hall Ltd).

5.22. DENNIS, N. T. M., and HEPPELL, T. A., 1968, *Vacuum System Design*, pp. 126 *et seq.* (Chapman & Hall Ltd).

5.23. POWER, B. D., 1966, *High Vacuum Pumping Equipment*, p. 397 (Chapman & Hall Ltd).

5.24. POWER, B. D., 1966, *High Vacuum Pumping Equipment*, Table 5.2, pp. 118–119 (Chapman & Hall Ltd).

5.25. MCILRAITH, A. H., and SCOTT, A. D. L., 1966, *Jour. Sci. Inst.*, **43**, p. 961.

5.26. DENNIS, N. T. M., and HEPPELL, T. A., 1968, *Vacuum System Design*, p. 183 (Chapman & Hall Ltd).

5.27. DENNIS, N. T. M., and HEPPELL, T. A., 1968, *Vacuum System Design*, pp. 190–191 (Chapman & Hall Ltd).

5.28. TURNBULL, A. H., BARTON, R. S., and RIVIÈRE, J. C., 1962, *An Introduction to Vacuum Technique*, pp. 173–175 (George Newnes Ltd).

5.29. KENT, T. B., 1955, *Jour. Sci. Inst.*, **32**, pp. 132–134.

5.30. TURNBULL, A. H., 1965, *Vacuum*, **15**, pp. 3–11.

5.31. TURNBULL, A. H., BARTON, R. S., and RIVIÈRE, J. C., 1962, *An Introduction to Vacuum Technique*, pp. 176 *et seq.* (George Newnes Ltd).

5.32. STEINHERZ, H. A., 1963, *Handbook of High Vacuum Engineering*. Chapter 11 (Reinhold Publishing Corporation, N.Y.)

Bibliographical Note

G. W. Green's book, *The Design and Construction of Small Vacuum Systems* (Chapman & Hall Ltd, 1968), will be found very helpful by novices in vacuum technique, who intend to set up a small laboratory system.

6

ULTRA-HIGH VACUUM

§ 6.1. General

If the definition of ultra-high vacuum is taken as a pressure of 10^{-9} torr or less, U.H.V. was probably achieved many years ago. However, this was not then recognized, because the soft X-ray effect in ionization gauges (see § 4.5.1) prevented the measured collector current from falling below a value which, interpreted in terms of pressure, corresponded to about 10^{-7} torr. However, Nottingham pointed out in 1947 that the real *positive ion* current was smaller than the measured collector current, since the latter was augmented by photo-electrons *leaving* the collector. The introduction of the Bayard-Alpert gauge (see § 4.5) allowed measurement down to 10^{-10} torr, and subsequent developments have extended the lower limit by several decades.

Once it became possible to measure low pressures, effort was directed to methods of obtaining them. Oil or mercury diffusion pumps equipped with L.N$_2$-cooled traps had ultimate pressures of 10^{-10} torr or below, but material outgassing prevented the attainment of U.H.V. in a work chamber in a reasonable pumping time. Hence, attention was concentrated on the materials and construction of chambers, and demountable seals bakeable to several hundred degrees centigrade, so that the room-temperature outgas rate after baking might be sufficiently small compared with available pump speed to enable U.H.V. to be attained reasonably quickly.

Thus U.H.V. technique is mainly concerned with the material and construction of work chambers and the measurement of low pressures attained in them. A useful general review of U.H.V. technique, with special reference to pumps and to total and partial pressure gauges, is given by Craig (ref. 6.1).

§ 6.2. Pumps

The production of U.H.V. can be achieved by several types of pump already described in Chap. 3. The subject has recently been treated in specialized works by Redhead, Hobson, and Kornelsen (ref. 6.2) and by Robinson (ref. 6.3), so only a few general remarks are made here.

Diffusion pumps with working fluids of mercury or oil can be used with success with well-designed cold traps, suitably prebaked. The advantage of mercury is that, being an element, it cannot decompose as maltreated organic pump fluids will. But a cold trap should preferably be interposed between the backing pump and the mercury pump, as well as between the latter and the work chamber, if certainty of exclusion of organic vapours is desired.

Since at a given temperature, the vapour pressure of pump oils is much lower than that of mercury, a mechanically refrigerated cold trap at $-40°C$ interposed between diffusion pump and work chamber allows (see ref. 6.4) pressures of about 10^{-10} torr to be attained. Use of mechanical refrigeration, rather than L.N_2, has considerable advantage in convenience.

A popular U.H.V. system consists of a sputter ion pump, initially evacuated to its starting pressure by a sorption pump (see § 3.5) and 'topped up' (i.e. pressure further reduced) by a titanium sublimation pump (see § 3.4.2). The latter need only be operated intermittently, as at low pressures a freshly sublimed layer of Ti remains active for a considerable time.

An interesting clean pump scheme for achieving $< 10^{-10}$ torr in a glass system of volume of about $\frac{1}{2}$ litre is described by Robinson (ref. 6.5). The system is originally filled with pure dry N_2 and then evacuated, via a cold trap, by a water jet pump to about 10 torr, and sealed off. The other pumps are molecular sieve sorption pumps, small Ti sublimation pumps, and an inverted magnetron gauge which also acts as a pump. These are operated and sealed off in succession, finally leaving the magnetron pump to deal with any He which may diffuse through the glass (see § 6.4) and the work chamber which is to be pumped.

§ 6.3. *Methods of pressure measurement*

6.3.1. *General*

The difficulty of U.H.V. measurement lies in the presence of spurious ion collector currents, due to causes such as the soft X-ray effect, rather than in the small value of the current itself. For example, in hot-cathode gauges with a sensitivity (see § 4.5) of 10 torr^{-1} and an ionizing electron current of 1 mA, the ion current corresponding to 10^{-12} torr is 10^{-14} A. However (ref. 6.6), currents of 10^{-15} A can be measured with a response time of ~ 1 sec, and much lower currents can be measured using secondary electron multiplier techniques.

6.3.2. *Hot-cathode gauges*

The lower pressure limit of the B.A.G. in its original form has been extended by Redhead (ref. 6.7) by modulation of the collector current. The essential feature is the introduction, within the grid space and parallel to the ion collector, of a wire whose potential can rapidly be changed from that of the grid (when it collects no positive ions) to that of the collector, when it reduces the collector current by some 30–40 per cent. This change, however, has little effect on the residual component of the collector current due to soft X-rays or other disturbing effects. The percentage modulation can be determined at pressures so high that the residual current is negligible compared with the true ion current. Using this modulation factor, the observed collector current at low pressures can be separated into true ion current (from which true pressure can be calculated) and the disturbing residual current.

Redhead has developed two other forms of U.H.V. gauge: the suppressor and the extractor types (ref. 6.8). In the suppressor gauge, the emission of photo-electrons from the ion collector is prevented by an electric field at its surface, which feature, combined with a modulator electrode, has allowed the measurement of pressures of about 10^{-14} torr.

In the extractor gauge the collector is a short fine wire. The ions pass through a small hole in a shield and are electrostatically focused on the collector, thus rendered partially immune from soft X-ray effects. The desorption of gases by electron bombardment of the grid is much less than in the B.A.G. or suppressor gauge. This is a decided advantage when measuring very low pressures. The lowest pressure measured by the gauge is 7×10^{-13} torr.

The orbitron gauge has already been mentioned (§ 4.5.2). It has a low X-ray limit and, since its sensitivity is so high, requires only a small electron current, and therefore temperature induced inter-action with the filament, and desorption from the anode under electron bombardment, are minimized.

In the bent-beam gauge developed by Helmer and Hayward (ref. 6.9) a beam of ions, extracted from a grid cage in which they are produced by electron bombardment, is bent through a 90° arc by an electric field and reaches a collector which cannot 'see' the grid cage. A Cary vibrating reed electrometer measures the ion current to the collector, photoelectric emission from which is prevented by a suppressor grid.

Hot-cathode low-pressure gauges have been developed in which a magnetic field both increases the electron path and suppresses the emission of photoelectrons from the ion collector surface. Details of two gauges of this type, developed by Lafferty and Klopfer, are given in ref. 6.10.

6.3.3. *Cold-cathode gauges*

Two types, having crossed electric and magnetic fields, are the magnetron and the inverted magnetron U.H.V. gauges. Both have cylindrical symmetry, the electric field being radial and the mag-netic field parallel to the axis, and of such magnitude that electrons emitted from the cathode and attracted to the anode are forced by the magnetic field (see Appendix II, § 16) to move in curved trajectories and return to the cathode, unless they lose energy by collision with a gas molecule.

In both types, the construction is such that only the ion current and not any field emission current (see Appendix II, § 14) is measured. The latter, however, is useful in starting the discharge at low pressures.

In the magnetron gauge, the cold cathode is an axial cylinder having two circular end discs and the anode is a coaxial squat cylinder. With an applied potential of 6 kV and an axial magnetic field of 1 kilogauss the sensitivity is of the order of 10 A torr^{-1}.

In the inverted magnetron (ref. 6.11), the anode is a cylinder and the cathode a coaxial cylinder surrounding it. The applied voltage is about the same as that of the magnetron gauge, but the axial magnetic field is 2 kilogauss.

According to Redhead (ref. 6.12), who has developed both types, the inverted magnetron gauge appears to be more stable and less noisy than the magnetron gauge.

The advantage of cold-cathode gauges for U.H.V. measurement is that, provided the geometry excludes the measurement of field emission electron current, the electron current is a function of the gas pressure. Hence, any soft X-ray effect produced by electrons striking the anode is also a function of gas pressure. Thus there is no soft X-ray limit.

§ 6.4. Construction and materials of U.H.V. systems

Much of the information already given in Chap. 5 is applicable to U.H.V. problems. The present section adds a few comments of special relevance.

A principal requirement of materials and couplings for U.H.V. is that they must be bakeable to several hundred degrees centigrade to reduce subsequent outgassing. Except for small nondemountable Pyrex apparatus, this in practice means vacuum envelopes of S.S. and avoidance within the work chamber of any metals with high vapour pressure. For example, hard solders, if used, must not contain zinc or cadmium.

If the vacuum chamber consists, as it almost invariably does, of a number of S.S. components bolted together, with metal/glass or metal/ceramic electrical lead-ins, these should be individually tested, before assembly, on a leak detector, preferably of the He-tuned mass-spectrometer type. It is good practice to bake out S.S.

parts in an auxiliary vacuum furnace before leak testing, in order to
'open up' any potential leaks, which can more easily be dealt with
at this stage.

To get some 'feel' for the magnitude of leaks 'real' and 'virtual',
consider a small stainless steel cylindrical chamber, diameter 10
cm and height 20 cm, and assume a real leak of 10^{-6} lusec, i.e.
10^{-9} torr litre sec^{-1}. The outgas rate of EN58B at room temperature, after a 16 hour bake at 400°C, is (ref. 6.13) 3×10^{-14} torr
litre sec^{-1} cm^{-2}. Hence the virtual leak would be only about $2\frac{1}{2}$ per
cent of the 'real' one, and to maintain a pressure of 10^{-10} torr the
required pump speed at the work chamber would be given by

$$S_c \times 10^{-10} = 1.024 \times 10^{-9}$$

or $$S_c \approx 10 \text{ litres sec}^{-1}$$

Leak consideration should include the fact that certain materials,
notably Pyrex and fused silica, are permeated by He and also take
it into solution. Since the partial pressure of He in the atmosphere
is about 4×10^{-3} torr, its presence may be pressure limiting in an
all-Pyrex apparatus.

In ref. 6.14, a most instructive calculation is given of the time
variation of He partial pressure in a spherical Pyrex vessel, of volume 1 litre and wall thickness 1·5 mm, evacuated by a 0·1 l sec^{-1}
pump. The attainment of a linear concentration gradient of
He in the Pyrex, and hence of a constant pressure of He in the
vessel, would take a long time at room temperature. After 10 hours'
pumping, the pressure would still be about one order greater than
the equilibrium, i.e. He would still be emitted from the inner
surface more quickly that it entered the outer one.

However a 10-hour bake at 500°C, during which the pressure
would rise to a steady value of 6×10^{-9} torr (owing to the increase
of the diffusion coefficient of He with temperature) would establish
a linear concentration gradient. This, because the *solubility* of He
in Pyrex is virtually temperature independent, would allow the
final equilibrium pressure of $8·5 \times 10^{-12}$ torr to be reached
immediately on return to room temperature.

The frictional forces between surfaces in contact is greatly increased in U.H.V., because the adsorbed layer of gas or other contaminant, which is normally present and acting as a lubricant, is absent or much reduced. Hence bearings and sliding surfaces easily seize, i.e. 'cold welding' takes place.

The effect (see e.g. refs. 6.15 and 6.16) can be minimized in several ways. One of these is the use of MoS_2 as a dry lubricant, though for really clean work the possiblity that dissociation may produce free S is disturbing. Another expedient is the use of a thin layer of some low-vapour-pressure element which is easily sheared and non-reactive, such as Au (ref. 6.17). A third method is the use, without intervening lubricant, of combinations of two metals found to have a small tendency to seize. Osborn (ref. 6.18) tried a number of bearing combinations of which a pure Fe shaft in a Ag bearing was found to be best.

References

6.1.   CRAIG, R. D., 1967, *Vacuum*, **16**, pp. 70–78.

6.2.   REDHEAD, P. A., HOBSON, J. P., and KORNELSEN, E. V., 1968, *The Physical Basis of Ultra-High Vacuum*, Chapter 11 (Chapman & Hall Ltd).

6.3.   ROBINSON, N. W., 1968, *The Physical Principles of Ultra-High Vacuum*, Chapter 2 (Chapman & Hall Ltd).

6.4.   TOLMIE, E. D., 1966, *J. Sci. Inst.*, **43**, pp. 954–957.

6.5.   ROBINSON, N. W., 1968, *The Physical Principles of Ultra-High Vacuum*, p. 71 (Chapman & Hall Ltd).

6.6.   REDHEAD, P. A., HOBSON, J. P., and KORNELSEN, E. V., 1968, *The Physical Basis of Ultra-High Vacuum*, p. 295 (Chapman & Hall Ltd).

6.7.   REDHEAD, P. A., HOBSON, J. P., and KORNELSEN, E. V., 1968, *The Physical Basis of Ultra-High Vacuum*, p. 312 (Chapman & Hall Ltd).

6.8.   REDHEAD, P. A., and HOBSON, J. P., 1965, *Brit. J. App. Phys.*, **16**, pp. 1555–1566.

6.9.   HELMER, J. C., and HAYWARD, W. H., 1966, *Rev. Sci. Inst.*, **37**, pp. 1652–1654.

6.10. REDHEAD, P. A., HOBSON, J. P., and KORNELSEN, E. V., 1968, *The Physical Basis of Ultra-High Vacuum*, pp. 319 *et seq.* (Chapman & Hall Ltd).

6.11. HOBSON, J. P., and REDHEAD, P. A., 1958, *Can. J. Phys.*, **36**, pp. 271–288.

6.12. REDHEAD, P. A., HOBSON, J. P., and KORNELSEN, E. V., 1968, *The Physical Basis of Ultra-High Vacuum*, p. 334 (Chapman & Hall Ltd).

6.13. POWER, B. D., 1966, *High Vacuum Pumping Equipment*, p. 404 (Chapman & Hall Ltd).

6.14. REDHEAD, P. A., HOBSON, J. P., and KORNELSEN, E. V., 1968, *The Physical Basis of Ultra-High Vacuum*, pp. 107 *et seq.* (Chapman & Hall Ltd).

6.15. WINSLOW, P. M., and McINTYRE, D. V., 1966, *J. Vac. Sci. and Tech.*, **3**, pp. 54–61.

6.16. ROWE, G. W., 1957, *Inst. of Mech. Eng. Proc. Conf. on Lubrication and Wear*, pp. 333–338.

6.17. SPALVINS, T., and BACKLEY, D. H., 1966, *J. Vac. Sci. and Tech.*, **3**, pp. 107–113.

6.18. CARPENTER, L. G., 1965, *Proc. Inst. Mech. Eng.*, **179**, Part 3G, pp. 165–171.

7

APPLICATIONS
OF
VACUUM TECHNOLOGY

§ 7.1. General

Vacuum technology was originally a tool of laboratory research used, for example, to remove the normal atmospheric gases so that the empty space (vacuum) could be filled with other gaseous species on which it was desired to experiment. A typical laboratory use was concerned with the investigation of electrical discharges through gases at low pressures, in which the mean free paths of atoms, ions, and electrons were large compared with their values at atmospheric pressures. The discovery of the electron, and the determination of its charge/mass ratio was made possible by this means.

Today, high vacuum has become such a common research tool that it would be difficult to find any sizeable research laboratory not possessing a number of pieces of vacuum-producing equipment. Thus, the present uses of vacuum in research are too numerous and varied to catalogue and, with the exception of Nimrod, will not be mentioned in this chapter.

However, on account of their economic importance and intrinsic interest, some industrial applications will be described briefly. These applications are mainly of two kinds.

One is concerned with the production, often in large numbers, of articles which are evacuated (e.g. thermionic valves and dewar flasks) or having been evacuated of normal atmospheric air are refilled with another gas or mixture of gases (as e.g. in gas-filled filament lamps or discharge lamps.)

The other main application is the manufacturing or processing plant, in which vacuum is used to produce articles which in their final form are not evacuated; examples range from vacuum-melted steel ingots (weights of order 200 tons) to evaporated electronic micro-circuits. For an excellent treatment of certain applications in more detail than the scale of the present book allows, see Pirani and Yarwood, ref. 7.34.

§ 7.2. Nimrod

This is a proton synchrotron designed to accelerate pulses of 10^{12} protons to an energy of 7 GeV (7×10^9 electron volts) at a rate of 28 pulses per minute.

The ions are accelerated to an energy of 15 MeV in an auxiliary apparatus (injector), evacuated to between 10^{-5} and 10^{-6} torr by mercury vapour pumps of total speed 10^4 litres sec^{-1}, whence they are injected into a circular vacuum vessel (mean diameter 48 metres) and constrained to move within it by a magnetic field perpendicular to the plane of the ring (see Appendix II, § 16). Once in each circuit of the ring, the energy of the ions is increased by 7 keV by means of a suitably phased r.f. electric field. They travel a distance of $1 \cdot 5 \times 10^5$ km before attaining their final energy of 7 GeV. To limit the loss by scattering collisions with gas molecules to an acceptably small value, the ring is evacuated to at least 10^{-6} torr, the lowest recorded pressure during commissioning being about 3×10^{-7} torr. In order to retain the ions within the ring as their speed increases (see Appendix II, § 15), the magnetic field is increased from about $0 \cdot 3$ to 14 kilogauss during the total acceleration time ($0 \cdot 7$ sec) of a pulse. The choice of material of the vacuum ring was governed by the following considerations. It had to be non-magnetic to avoid distorting the carefully designed magnetic field, and since this field varies during the orbiting of a pulse the material had to be sufficiently non-conducting to avoid induced currents which would again disturb the magnetic field. In addition, the material had to be resistant to ionizing radiations produced in the ring and have an outgas rate low enough to allow the attainment of the necessary vacuum. The material finally chosen was glass-fibre-reinforced resin, double walled within the magnetic gap. The outer vessel was evacuated to about 1 torr and supported by the pole tips, to prevent collapse under atmospheric pressure, and the inner vessel, which therefore had only to support its own weight, was evacuated by forty 60 cm diameter oil-diffusion fractionating pumps, with a total speed of 10^5 litres sec^{-1} at 10^{-6} torr. The backing pumps were of the vapour booster type, themselves backed by small rotary pumps. The inner vessel had a stainless steel lining, comprising 99 per cent of its area, applied in strips to limit eddy currents, and having an outgassing rate (see § 5.8) of about 10^{-9} torr litre cm^{-2} sec^{-1}, giving an average outgassing for the whole inner area of about 10^{-8} torr litre cm^{-2} sec^{-1}. For full details of this large-scale example of vacuum engineering, see ref. 7.1 by G. S. Grossart, to whom I am indebted for helpful personal communications.

§ 7.3. High vacuum as an electric insulator

Variable capacitors, in which the dielectric is high vacuum may be used in substitution for air dielectric capacitors geometrically many times larger, because the breakdown voltage at a given electrode spacing is much higher *in vacuo* than in air.

In the type manufactured for r.f. use by the English Electric Valve Co. Ltd (to whose courtesy I am indebted for technical information) the electrodes consist of two sets of interpenetrating coaxial metal cylinders. The vacuum envelope is partly ceramic or glass and partly bronze bellows, which latter allow the degree of interpenetration to be varied by axial movement of one set of cylinders (i.e. one electrode); hence the capacity is variable but vacuum integrity preserved. Manufacture includes outgassing by heat during evacuation by diffusion pumps. Getters are not used, since any films of deposited material tend to increase the possibility of field emission (see Appendix II, § 14), which in turn leads to premature breakdown. Capacitors are available with rated peak voltages between 3 and 30 kV and the possible range of variable capacity is of the order 10 to 10^3 picofarads. The degree of vacuum is 10^{-7} to 10^{-8} torr and the nominal clearance between adjacent cylinders is 1 mm per 10 kV of rated working peak r.f. voltage.

Vacuum relays with rated operation voltages up to 50 kV (D.C.) are commercially available (ref. 7.2) with a maximum operating time of 50 ms and a maximum stable contact resistance of 0·015 Ω.

The successful testing of a 132 kV vacuum circuit breaker has been announced (ref. 7.3). The vacuum interrupters are self-contained within an evacuated envelope, and the moving contacts, operated through flexible bellows, are moved only a short distance to achieve interruption. The use of vacuum interrupters for switching and circuit breaking has been reviewed by Reece (ref. 7.4).

§ 7.4. Separation by vacuum distillation

While in many cases the industrial separation by distillation of the components of a parent substance composed of several species of different vapour pressure is carried out at atmospheric pressure, vacuum distillation is in some situations advantageous or essential.

The presence of the normal atmosphere impedes the passage of molecules from the heated parent substance to the condenser, since the mean free path at 760 torr (10^{-5} cm) involves many scattering collisions in transit. Hence reduction of pressure increases distillation speed.* Moreover, the latent heat of evaporation at atmospheric pressure exceeds that *in vacuo* by the amount of work done in expansion against the atmosphere. Hence the thermal energy requirement is less in the latter case and the same distillation rate can be achieved at a lower temperature.

For large organic molecules, there is an upper molecular weight limit above which the energy needed for conversion from the

---

* The reverse process is exemplified by the use of inert gas in gas-filled W-filament lamps to impede filament evaporation.

solid to the liquid phases is of the same order as that required to cause molecular breakup. Generally this limit is at a molecular weight of 200 to 300 or boiling point of 250° to 300°C.

The extraction of vitamins from parent substances such as fish liver or vegetable oil, described in ref. 7.5, exemplifies high vacuum unobstructed path distillation on a commercial scale. The removal of atmospheric gases to a residual pressure level of order 10^{-4} torr enables distillation to be carried on at an economic rate at temperature of order 200°C, thus avoiding higher temperatures which would cause thermal degradation of the distillate. This is the fundamental advantage of vacuum distillation in this application. The low number density of the residual atmospheric molecules also virtually eliminates damaging oxidative collisions. However the vapour pressure of the distillate molecules themselves can be much higher (order 10^{-2} torr) in the interest of high distillation speed, since collisions between them are not generally harmful.

For detailed information on molecular stills, the monograph (ref. 7.6) by P. Ridgway Watt, whose help I gratefully acknowledge, should be consulted.

§ 7.5. Freeze drying

The object of this is to remove water from organic substances (e.g. from food or from skin and arteries for surgical grafting) in such a way that the material can be stored at room temperature for many months and reconstituted in approximately its original condition by the addition of water.

In essence a freeze dryer consists of a vacuum chamber in which pre-frozen material is placed. The partial pressure of air is reduced to between 10^{-1} and 10^{-2} torr by a pump, and the sublimed water vapour removed by means of what is essentially a cryo pump, namely a refrigerated condenser with a surface temperature of about -40°C. The partial pressure of air is usually about a tenth of that of the water vapour. Hence the passage of the latter to the condenser is not hindered.

When wet solids containing dissolved or suspended substances are dried, the removal of water often translocates them—a process not reversed when the substance is re-wetted. In freeze drying on the other hand, the water sublimes from the solid phase (ice) and translocation is reduced or eliminated. Since the sublimation of a gram of water needs about 700 calories, it is necessary to add heat by radiation or conduction to the product while it is being freeze dried.

The plant requirements and economics of the two main classes of freeze dried products (edible and biological) have been compared by Longmore (ref. 7.7). Process cost is a major consideration only in the case of food products. For biological products, the important criteria are sterility, stability, and minimum chemical and biological change, and the plants are pumped by refrigerated condensers and oil-sealed mechanical pumps. Large amounts of water vapour are sometimes removed by steam ejector pumps. Ref. 7.31 describes the freeze drying of human blood plasma. Most of the water is removed under vacuum by a refrigerated condenser, and the last portion by vacuum drying over P_2O_5.

In industrial-scale food plants operating a number of units simultaneously, the refrigerated condensers are often served by a common central refrigerating system. The principles of the freeze-drying of food have been reviewed by Rowe (ref. 7.8) and further information is contained in ref. 7.33. An example of freeze drying in the food industry is the large-scale production of a well-known brand of 'instant' coffee by this means.

§ 7.6. Manufacture of lamps and thermionic devices

One requirement in lamp manufacture is the attainment of an acceptable filament life, necessitating the removal of gases such as H_2O which interact chemically with the W, and the introduction of inert gases (Ar and N_2) which will decrease its sublimation rate. The lamps are alternately evacuated by a rotary pump and flushed with N_2 and finally sealed off, when containing a mixture of N_2 and Ar at about one atmosphere pressure. Final traces of reactive gas are removed by applying the normal operating voltage which fires a red phosphorus getter.

8

In thermionic valve manufacture, it is imperative to avoid small partial pressures of gases which may 'poison' the emission of the oxide-coated filament. Evacuation to about 10^{-5} torr by diffusion and rotary pump is followed by the firing of a getter which reduces the pressure by about another three decades. The vacuum technique of cathode-ray-tube manufacture is similar. Television tubes are often pumped by diffusion pumps, backed by rotary mechanical pumps. Silicone oils are used in the former since they will stand exposure to atmosphere when still hot—an important feature for quick cycling. Details of a typical pumping group for final-pumping of television tubes are given by Power (ref. 7.9). Ward and Bunn (ref. 7.32) give an account of the application of vacuum technique to the manufacture of electron tubes and lamps.

§ 7.7. Vacuum metallurgy

Vacuum technology is of considerable and increasing interest in metallurgy, because many metallurgical processes involve heat, and most metals when heated react with atmospheric air.

Vacuum metallurgical processes can be considered under two main heads—the removal of dissolved gases or other unwanted species from metals, and the carrying out of operations which the presence of air would impede or prevent.

The reduction of the hydrogen content of steels by vacuum treatment is of increasing importance. Very small concentrations of dissolved hydrogen in high tensile steel can lead to embrittlement (ref. 7.10) and vacuum degassing in the molten state produces significant improvement. In 1968, more than 250 plants were in operation, in some of which ingots up to 250 tons could be cast. The treatment is also used for the controlled de-oxidation of steel melts and permits the introduction of reactive alloying elements which would otherwise be partly and uncontrollably lost by oxidation. For details, and other information re vacuum melting, see ref. 7.11.

Electron-beam guns, developing powers of several hundred kilowatts per gun, have important metallurgical uses, especially in the melting, in water-cooled metal crucibles, of refractory metals such as Ta at temperatures so high that reaction with conventional ceramic crucibles would produce unacceptable contamination (ref. 7.12). Large-scale electron-beam processes in melting steel degassing, slab conditioning, annealing and heat treatment, and evaporative coating are reviewed in ref. 7.13.

Electron-beam welding (a specialized application of electron-beam guns) is characterized by extreme power *density*, good controllability, and the possibility of welding in cavities of a large depth-to-width ratio. Although originally developed for small-scale work, where very precisely located welds are required, it is by no means confined to these applications, and it has recently been reported (ref. 7.14) as being used in the USA by the Ford Motor Co. for the mass production of relatively large components, i.e. starter flywheel assemblies for cars, which were formerly produced by conventional welding methods.

In the original form of the electron gun, the workpiece to be welded had to be introduced into the vacuum chamber (pressure of order 10^{-4} torr), in which the electron gun itself operated—a serious limitation on speed of production and on size of workpiece. This limitation has however (ref. 7.15) been overcome and the workpiece itself may be at atmospheric pressure. Electron-beam welding of electronic components is discussed in ref. 7.16.

Vacuum brazing (pressure $\sim 10^{-4}$ torr) is useful for components small enough to go into a vacuum furnace. It is fluxless, gives good and reliable joints (ref. 7.17) but is generally higher in capital cost of plant and lower in output than a H_2 furnace of the same charge size. However it has definite advantages for the brazing of complex structures with interstices, such as heat exchangers, which are difficult to flush adequately with H_2. The advantages, technical and economic, are discussed in ref. 7.18.

§ 7.8. Coating by vacuum evaporation

In some optical instruments, e.g. cameras, the fact that each air/glass surface reflects about 5 per cent of the incident light may be serious. This can be considerably reduced by 'blooming', i.e. by coating the glass surface with a transparent medium of suitable refractive index. Magnesium fluoride or SiO are often used, a thickness of order 10^{-5} cm being applied by vacuum evaporation (ref. 7.19) from a silica or Ta crucible.

In micro-electronics (ref. 7.20), the passive elements of thin-film integrated circuits (i.e. the resistors and capacitors) are fabricated by vacuum deposition in background pressures of order 10^{-5} torr or lower. Ni-Cr of thickness of order 10^{-6} cm is much used for resistors, and SiO as the dielectric of capacitors; the electrodes and interconnections are often of evaporated Al.

Tantalum, which is a high-melting-point metal and difficult to evaporate, has been sputtered (see Appendix II, § 18) in pure dry Ar to form resistors and, anodically oxidized, used as the dielectric of capacitors.

The use of vacuum-deposited thin-film micro-circuits from the point of practical manufacture is reviewed by Gaffee (ref. 7.21) and their use in silicon-chip semi-conductor integrated circuits by Glang (ref. 7.22).

Aluminizing of the front surface of mirrors is carried out in pressures of about 10^{-5} torr (ref. 7.23), producing a maximum reflectance for visible light of about 92 per cent. The resistance to tarnish and corrosion is excellent and the mechanical strength reasonable. A layer of SiO is sometimes deposited on the aluminium to give additional protection against abrasion.

Steel is often protected against corrosion by a thin coat of Cd, applied by electrolysis, which however also introduces hydrogen, causing loss of strength if the steel is of the high tensile type (ref. 7.10). Vacuum deposition of Cd carried out under proper conditions of surface cleanness gives adhesion as good as that obtained by electrodeposition but avoiding hydrogen embrittlement (ref. 7.24).

Plant for the continuous production of steel strip protected on one or both sides by evaporated metal (e.g. Al) has operated successfully on a large scale (ref. 7.25). The strip enters the plant at atmospheric pressure, and is led via a number of thin slit apertures through a series of chambers pumped to successively lower pressures into the evaporation chamber at 10^{-4} torr, and out again to the atmosphere through similar slit aperture systems. It is claimed that this process uses less electric power, per unit of material deposited, than electrolytic plating.

Vacuum coating, unlike electro-plating can be applied to electrical insulators. Hence plastics also are coated on an industrial scale by vacuum evaporation (ref. 7.26). First a coating of lacquer is applied to limit the degassing of the plastic base and to produce a smooth surface for the following evaporated metal coating—often aluminium of thickness of order 10^{-5} cm. The final stage is the application of an outer protective coat of lacquer (which may if desired be dyed) producing a smooth lustrous decorative finish. The process is used in the automobile, artificial jewelry, and toy industries.

However, the vacuum metallizing of plastics has special problems, arising from degassing which despite the first lacquer coat exceeds that of glass by several orders of magnitude. As water or condensible vapours are a major constituent of the outgassing, cryo pumping can be usefully employed in conjunction with more conventional equipment.

Holland and Barker (ref. 7.27) have described one such system in which a 72-inch-diameter work-chamber 72 inches long is pumped by six combined diffusion/booster pumps (Edwards type 9B4) having a combined speed for air (when baffled) of 9000 litres sec^{-1}, and backed by a 125 litre sec^{-1} gas-ballasted rotary pump. The cryo pump, cooled by L.N_2, is in the form of a flat helix of copper tubing having a calculated pump speed for H_2O vapour of about 2×10^5 litres sec^{-1}. The vessel contains a rotating work-holder and, under a plastic load of 1200 acrylic mouldings (each 1 oz in weight), a vacuum suitable for metal finishing (viz. 5×10^{-4} torr) can be attained in 12 minutes if the cryo pump is used, or in 19 minutes without it.

§ 7.9. Environmental testing of spacecraft

The malfunctioning of a spacecraft in service is extremely wasteful and, if the craft is manned, may be fatal. Hence the necessity for extensive terrestrial environmental testing, which includes testing *in vacuo*.

The major effects of vacuum on spacecraft include the rapid pressure reduction during launch, the virtual elimination of gaseous thermal conduction in orbit, the volatilization of materials, or some components of them, and effects of vacuum on the function of bearings.

The effects of the launch conditions on corona discharge in electrical apparatus and mechanical stresses between the interior of the vehicle and its environment are discussed by Lorenz (ref. 7.28), who gives details of pumping equipment used to produce a pressure/time profile which reproduces that of the launch.

In orbit, the thermal state of a space vehicle is determined by the balance between solar radiation received, the radiation it emits to space (effective temperature about $4°K$), its own internal thermal linkages, and its own power dissipation—which is usually small. At pressures below about 10^{-5} torr, thermal linkage due to gaseous conduction may be neglected (ref. 7.29), so that as far as thermal testing is concerned this is a sufficiently low internal pressure. Hence, the requirements for thermal-balance test chambers are a pressure external to the spacecraft of $\sim 10^{-6}$ torr and an internal pressure $\ngtr 10^{-5}$ torr, solar simulation of the order of 1 solar constant,* and a heat sink at a temperature such that the radiation from it to the space vehicle is small compared with that emitted by the vehicle itself. Since radiation is approximately proportional to the fourth power of the absolute temperature, the heat sink need not be cooler than $100°K$, a temperature conveniently produced by an $L.N_2$-cooled shroud. This also serves as a cryo pump for the easily condensed gases and supplements the oil-diffusion pumps usually employed.

Excessive outgassing of materials in space vehicles may make the time to attain the necessary vacuum for internal thermal tests

* The solar constant is the amount of solar energy which (in the absence of atmospheric absorption) would fall in unit time on unit area of a surface at the mean distance of the earth from the sun, and so placed that the solar radiation falls normally upon it. Its value is approximately 0.13 watt cm^{-2}.

($\sim 10^{-5}$ torr) intolerably long. It may also be detrimental if material sublimes from the warmer to the cooler parts, where its presence affects function, or if the loss of volatile constituents degrades mechanical properties, e.g. loss of flexibility in plastics. While a vacuum of order 10^{-5} to 10^{-6} torr is sufficient for thermal testing and to allow sublimation to take place unimpeded (ref. 7.29), it is not adequate for the testing of bearings and sliding surfaces.

In space, the pressures are in the U.H.V. region (e.g. at 600 km the particle density corresponds to an air pressure at laboratory temperature of $\sim 10^{-10}$ torr); in interplanetary space, they are many decades lower. Long exposure may remove the adsorbed layers from surfaces not in pressurized compartments, and increased friction and seizure (see § 6.3) may result. Hence such components should be additionally and individually tested in small U.H.V. chambers, since outgassing of the complete space vehicle makes the attainment of realistic vacua impracticable in large environmental chambers.

There are in existence a number of space-simulation chambers in which full-size satellites can be tested. In Great Britain, the largest is that at the Royal Aircraft Establishment, at Farnborough, Hampshire (ref. 7.30). The chamber is large enough to accept any spacecraft which can be included within a 2·5 metre diameter sphere, and solar simulation is achieved by six carbon arc lamps shining through six 20 cm diameter ports, capable of giving the equivalent of about $1\frac{1}{2}$ solar constant on a 2·5 metre diameter spacecraft. The pumping system consists of two 91·5 cm diameter oil-diffusion pumps backed by a rotary pump of capacity 8500 litres min^{-1}. The chamber can be pumped down to below 10^{-6} torr, at which the effective pumping speed is at least 25 000 litres sec^{-1}. A L.N$_2$-cooled liner aids the pumping of condensible vapours.

§ 7.10. *Vacuum as thermal insulation*

When the pressure in an evacuated space is reduced to a value such that the average distance of unimpeded motion of the gas molecules is limited by their collisions with *solid* obstructions (as opposed to other gas molecules), the effective thermal conductivity falls below the value given by equation 10 of Appendix I. This principle is applied in the domestic picnic flask and on a large scale in industrial containers for liquefied gases, such as L.N$_2$ and L.O$_2$. Typically, the thermal insulation of these large industrial containers consists of an evacuated annular space, filled with a heat-insulating powder, the average size of the interstices of which are small compared with the mean free path appropriate to gas/gas collisions at this pressure (see Appendix I, § 9). Thus the condition for reduced thermal conductivity is fulfilled.

The powder particles also form a series of radiation shields between the outer vessel at room temperature and the inner one at the temperature of the liquefied gas.

One material of construction is aluminium and the powder-filled space is evacuated via a cold trap (§ 2.6) by a rotary mechanical pump, possibly augmented by a diffusion pump. The service pressure, after sealing off, is 2×10^{-1} torr when the inner vessel is empty and 10^{-2} torr when it is filled with liquefied gas, the reduction being due to sorption pumping (see § 3.5).

I am indebted to the courtesy of the British Oxygen Company for technical information. For further information ref. 7.35 (which itself contains twenty-two references) should be consulted.

References

7.1.   GROSSART, G. S., 1966, *Trans. Third Internat. Vac. Congr.*, Vol. 1, pp. 89–110 (Pergamon Press Ltd).

7.2.   Marketed in Great Britain by Walmore Electronics Ltd.

7.3.   AEI Ltd, *Publication*, 1305–1371, Ed. A(1167).

7.4.   REECE, M. P., 1968, *Proc. Third Internat. Symp. on Discharges and Electrical Insulation in Vacuum*, Paris, pp. 305–314.

7.5.   WATT, P. R., 1961, *Proc. Symp. on User Experience of Large Scale Industrial Vacuum Plant*, p. 107 (Inst. of Mech. Eng.).

7.6.   WATT, P. RIDGWAY, 1963, *Molecular Stills* (Chapman & Hall Ltd).

7.7.   LONGMORE, A. P., 1968, *Proc. Fourth Internat. Vac. Congr.*, Part 1, pp. 79–82 (Institute of Physics and Physical Society, London).

7.8. ROWE, T. W. G., 1963, *Inst. of Fuel Repub. of Ireland Group, Conf. on Drying*, Dublin.

7.9. POWER, B. D., 1966, *High Vacuum Pumping Equipment*, p. 112 (Chapman & Hall Ltd).

7.10. ROLLASON, E. C., 1961, *Metallurgy for Engineers*, 3rd Edition, p. 216 (Edward Arnold (Publishers) Ltd).

7.11. WINKLER, O., 1968, *Proc. Fourth Internat. Vac. Congr. Part 2*, pp. 454 *et seq.* (Institute of Physics and Physical Society, London).

7.12. DIETRICH, W., GRUBER, H., SPERNER, F., and STEPHAN, H., 1968, *Vacuum*, **18**, pp. 657–663.

7.13. COAD, B. C., HUNT, C. d'A, SMITH, R. H., 1966, *Sheet Metal Industries*, **43**, pp. 800–807.

7.14. *Electrical News for Industry*, Winter 1968/9 (Electricity Council Marketing Department).

7.15. BAKISH, R., 1968, *Proc. Fourth Internat. Vac. Congr. Part 2*, p. 464 (Institute of Physics and Physical Society, London).

7.16. SANDERSON, A., 1967, *Electronic Components*, October, pp. 1135–1142.

7.17. HARVEY, J. T. A., and PERRY, E. R., 1963, *Welding and Metal Fabrication*, February.

7.18. BOSTON, M. E., 1965, *Machinery*, **106**, pp. 1079–1083.

7.19. DITCHBURN, R. W., 1963, *Light*, Vol. 1, pp. 138 *et seq.* (Blackie and Son Ltd).

7.20. HOLLAND, L., 1965, Editor, *Thin Film Micro-Electronics* (Chapman & Hall Ltd).

7.21. GAFFEE, D. I., 1965, *Thin Film Micro-Electronics*, Chapter 6 (Chapman & Hall Ltd).

7.22. GLANG, R., 1966, *Jour. Vac. Sci. and Tech.*, **3**, pp. 37–47.

7.23. HOLLAND, L. and BARKER, D. W., 1965. *Vacuum*, **15**, p. 290.

7.24. Patent Specification No. 43032/63, Inventors D. CLARK, and L. G. CARPENTER.

7.25. DIETRICH, W., HAUFF, A., and REICHELT, W., 1968, *Proc. Fourth Internat. Vac. Cong.—Part 2*, pp. 573–578 (Institute of Physics and Physical Society, London).

7.26. BARKER, D. W., 1963, *Internat. Plastics Engineering*, **3**, pp. 46–52 and 116–119.

7.27. HOLLAND, L. and BARKER, D. W., 1965, *Vacuum*, **15**, pp. 289–299.

7.28. LORENZ, A., 1968, *Proc. Fourth Internat. Vac. Congr.—Part 1*, pp. 213–218 (Institute of Physics and Physical Society, London).

7.29. CARPENTER, L. G., 1965, *Proc. Inst. Mech. Engrs.*, **179** 3G, p. 165.

7.30. EARL, A. G., SWIFT, R. D., and SPOONER, A. H., 1968, *Spaceflight*, **10**, pp. 174–182.
7.31. MILNE, G. R., and PETRIE, D. S., 1961, *Proc. Symp. User Experience of Large Scale Industrial Vacuum Plant*, pp. 99–106 (Inst. Mech. Eng.).
7.32. WARD, L., and BUNN, J. P., 1967, *Introduction to the Theory and Practice of High Vacuum Technology*, pp. 181 *et seq*. (Butterworth and Co. (Publishers) Ltd).
7.33. COTSON, S., and SMITH, D. B., 1963, *Freeze-Drying Foodstuffs* (Colombine Press (Publishers) Ltd).
7.34. PIRANI, M., and YARWOOD, J., 1961, *Principles of Vacuum Engineering*, Chapters 10–14 (Chapman & Hall Ltd).
7.35. *Cryogenic Engineering News*, 1969, February, pp. 18–21.

Appendix I

Some Relevant Formulae
in the
Kinetic Theory of Gases

§ 1. *The Maxwellian distribution of speeds*

If $f(c)\ dc$ is the fraction of molecules having speeds between c and $c + dc$,

$$f(c) = \frac{4}{\pi^{1/2}}\left(\frac{m}{2kT}\right)^{3/2} c^2\, e^{-mc^2/2kT} \tag{1}$$

where m = mass of molecule
 T = absolute temperature
 c = speed
 k = Boltzmann's constant = $1 \cdot 38 \times 10^{-16}$ erg deg^{-1}

From equation (1) can be deduced $\bar{c}$ (equation 3 below) and also $\bar{c}^2$ (equation 2), and hence the average kinetic energy of the whole assembly of molecules (equation 4). When $f(c)$ is plotted against c, it has a maximum at a speed called *the most probable speed C* [which is $(2kT/m)^{1/2}$, and hence 88 per cent of $\bar{c}$] and falls to negligible values at very high and very low speeds. An idea of the *breadth* of the distribution is given by the statement that 87 per cent of the molecules have speeds between twice and half C.

§ 2. *The mean-square speed*

$$\overline{c^2} = \int_0^\infty f(c)c^2\, dc$$

$$= \frac{3kT}{m} \tag{2}$$

§ 3. *The mean speed*

$$\bar{c} = \int_0^\infty f(c)c\, dc$$

$$= \left(\frac{8kT}{\pi m}\right)^{1/2} \tag{3}$$

§ **4.** *The average kinetic energy*

$$K.E. = \tfrac{1}{2}m\overline{c^2}$$
$$= \tfrac{3}{2}kT \text{ [from equation (2)]} \qquad (4)$$

§ **5.** *Equation of perfect gases*

$$P = \tfrac{1}{3}nm\overline{c^2}$$

where P is the pressure and n the number of molecules per cm^3

$$= nkT \quad \text{[from equation (4)]} \qquad (5)$$

§ **6.** *The average rate of bombardment of a surface*

$$\beta = \tfrac{1}{4}n\bar{c} \text{ molecules cm}^{-2}\text{ sec}^{-1} \qquad (6)$$

Substitution from equations (3) and (5) for $\bar{c}$ and n gives

$$\beta = P\left(\frac{1}{2\pi mkT}\right)^{1/2} \qquad (7)$$

§ **7.** *Quantities of gas*

From equation (5)

$$PV = nVkT \qquad (8)$$
$$= \text{total number of molecules in } V \times kT$$

where V is the volume considered. Hence, at a standard temperature—say 25°C (298°K) PV is a measure of the total number of molecules in the volume V. So, quantities of gas are quoted in units of pressure × volume (commonly called 'PV Units'), e.g. in torr litres, or micron litres (1 micron = 10^{-3} torr). At 25°C 1 micron litre contains $3 \cdot 24 \times 10^{16}$ molecules or 1 torr litre contains $3 \cdot 24 \times 10^{19}$ molecules.

§ **8.** *Flow rates*

These are often expressed in units of pressure × volume per unit time, e.g. the specific outgassing rate of a surface per unit area is given in torr litre sec^{-1} cm^{-2}. It is tacitly assumed, unless otherwise stated, that when quantities of gas are expressed in PV units, the temperature is that of the laboratory—say 298°K or thereabouts.

Leak rates are often expressed in torr litre sec^{-1} or micron litre sec^{-1}. A common unit is the lusec, which is a leak rate of 1 micron litre sec^{-1} and, at 25°C, is $3 \cdot 24 \times 10^{16}$ molecules sec^{-1}.

§ 9. *Mean free path* λ

The mean free path λ is the average distance travelled by a gas molecule before colliding *with another gas molecule*, and is approximately given by

$$\lambda = \frac{1}{n\pi\sigma^2} \qquad (9)$$

where σ is the molecular diameter, which does not vary greatly between gases, being of order 10^{-8} cm, unless the molecule is a large organic one.

Since $P = nkT$ (equation 5), at a given temperature λ varies as P^{-1} and, at room temperature, a useful rough rule is that λ is inversely proportional to P, being about 5 cm at 10^{-3} torr.

§ 10. *The thermal conductivity* K *of a gas*

$$K \approx \tfrac{1}{3}m\lambda n\bar{c}C_v \qquad (10)$$

where C_v is the specific heat at constant volume. Hence, at constant temperature ($\bar{c}$ constant), K is independent of n (i.e. of P) as long as $n\lambda$ is constant. When, however, P is reduced to the value where λ is greater than the vessel dimensions and the average free path no longer increases as n is diminished, the effective thermal conductivity decreases with P. Hence, for example, the thermal insulation of the evacuated annular space in a Dewar flask.

APPENDIX II

Notes on Topics in Physics
Relevant to Vacuum Technology

§ 1. *Atomic and molecular magnitudes*

Atoms consist of nuclei of diameter about 10^{-12} cm, surrounded by electron clouds of diameter about 3×10^{-8} cm. So molecules (except large organic ones) can be considered as having a target area of 10^{-15} cm^2 and a surface *monolayer* consists of about 10^{15} molecules cm^{-2}.

In a gas at N.T.P., the average spacing between centres of molecules is ~ 10 diameter as opposed to about one diameter (i.e. touching) as in solids and liquids.

§ 2. *Units and constants*

$$\text{Charge of electron} = 1\cdot6 \times 10^{-19} \text{ coulombs}$$
$$\text{1 electron volt} = \text{a unit of energy}$$
$$= 1\cdot9 \times 10^{-19} \text{ joules}$$

A *mole* is an assembly of atoms or molecules whose mass is equal to the atomic or molecular weight of the substance in question, e.g. 1 mole of H weighs 1 gram, 1 mole of H_2 weighs 2 grams, one mole of O_2 weighs 32 grams.

§ 3. *Evaporation and vapour pressure*

Consider a substance placed in an evacuated isothermal vessel (temperature $T°$K) with walls which do not react with it.

Under the influence of thermal agitation at temperature T, molecules will be emitted by the substance until a pressure P is built up in the vessel, such that the number of molecules striking and sticking to the substance per unit time per unit area is equal to the number emitted from it. P is called *the saturated vapour pressure* of the substance at the temperature T.

P is a strongly increasing function of T. For example, the vapour pressure of H_2O is 17·5 torr at 20°C and 760 torr at 100°C.

If the conditions are such that the number of molecules returning and sticking to the substance is less than the number emitted, it is said to evaporate or sublime, the former term applying to a liquid and the latter to a solid. However, there is no *essential* difference between the two processes.

The great temperature sensitivity of sublimation is exemplified, for example, by Zn. At $300°K$, the rate of recession of the surface would be about one monolayer ($\sim 10^{-8}$ cm) per year, the rate at $600°K$ being about 4 metres per year if sublimation were unimpeded.

§ 4. *Outgassing*

The emission into the gas phase of a foreign substance from the surface of a solid is called outgassing or desorption. An important example is the emission of H_2O vapour from the surface of glass or metal under vacuum conditions. This, like sublimation, is very temperature dependent, hence the common practice of heating a vacuum vessel, so that its rate of outgassing is much decreased *when it cools down again*. A vessel whose walls or contents are emitting gas is often said to have a *virtual leak*.

§ 5. *Adsorption*

This is the reverse of outgassing, being the process of capture by mutual intermolecular attractive force, on impact of gas molecule with adsorbing surfaces. The average probability that an impact will result in capture is called the 'sticking coefficient'. The degree of adsorption coverage under a given set of conditions is determined by the competition between the attractive forces (favouring adsorption), and the thermal motion of the molecules of the solid, tending to cause sublimation.

The amount of adsorption is often specified as the number of monolayers, fractional or integral, of the adsorbed vapour which adheres to the solid substrate.

The adsorbed material is referred to as the *adsorbate* and the substrate as the *adsorbent*.

When the number of adsorbed molecular layers exceeds (say) three or four, the adsorbed layer is indistinguishable from the adsorbed material *in bulk,* and may be called a macroscopic film in equilibrium with the normal vapour pressure at that temperature. When the film is a monolayer or less, the vapour pressure with which it is in equilibrium is less than that of the adsorbate in bulk.

Since, in adsorption, the mutual attractive forces between adsorbate and adsorbent do work on the adsorbate molecule as it approaches, there is a release of energy as the adsorbate 'sits down' on the adsorbent. This energy is called the *heat of adsorption,* and varies for different adsorption processes of interest in vacuum technology from about $\frac{1}{2}$ to $\frac{1}{20}$ electron volts (see § 2) per molecule. There are 6×10^{23} molecules in a mole, and heats of adsorption are often expressed in kilo-calories per mole, 1 electron volt per molecule being equivalent to 23 kilo-calories per mole.

§ 6. *Thermal accommodation coefficient*

When gas molecules at temperature T_g hit and rebound from a hot solid at temperature T_s, they acquire an amount of energy which depends on mutual intermolecular forces during impact and on the temperatures T_g and T_s. If the average energy, after impact, of the impinging molecules can be characterized by a temperature T_g', the accommodation coefficient α is defined by

$$\alpha = \frac{T_g' - T_g}{T_s - T_g} \tag{1}$$

i.e. by the ratio of the energy actually gained on impact to the maximum which could have been gained, if the impinging molecules had acquired the full temperature of the solid.

Efficiency of energy transfer depends on mutual interaction between impinging and surface molecules, and hence on adsorption. For many adsorption-laden surfaces $\alpha \approx 1$, but e.g. on *clean* W at 2800°K, $\alpha < 10^{-1}$ (Carpenter, L. G., Humphries, D. E., and Mair, W. N., 1963. *Nature,* **199,** p. 134).

§ 7. *Gettering*

This is a term used to describe the *irreversible* adsorption of a gas by a solid with which it combines to form a *chemical* bond; e.g. O_2 and N_2 are gettered by hot Ti.

§ 8. *Diffusion*

Diffusion is a process in which particles (atoms or molecules) move through a solid, liquid, or gas by means of a *random walk* mechanism, executed in virtue of their thermal motion.

In gas at pressures at which inter-molecular collisions are much more frequent than collisions with the walls of the vessel, the length of each random excursion is controlled by collision with neighbouring particles, and the average length of such excursion is called the *mean free path* (see Appendix I, § 9).

In vacuum situations, in which the mean free path is limited by collisions with the walls of the vessel, rather than by collisions with other molecules, the diffusion coefficient, as usually defined in the kinetic theory of gases, is not of much interest. However, the diffusion coefficient of gases *in solids* IS relevant to vacuum technology, since here the mean free path is fixed by collision with particles of the solid, and is of order 10^{-8} cm—i.e. the spacing of atoms in solids (Appendix II, § 1). The diffusion coefficient D can be defined in a number of ways which are equivalent. A useful definition for our present purpose is as follows. Consider a group of particles diffusing down the z axis, and all starting from $z = 0$ at $t = 0$. Then, at time t, the mean square distance of the particles from the origin is given by

$$\overline{z^2} = 2Dt \tag{2}$$

D, like vaporization rates, is a strong function of T. For example, D for H_2 in a certain stainless steel is $5 \cdot 0 \times 10^{-14}$ cm^2 sec^{-1} at room temperature and $3 \cdot 5 \times 10^{-8}$ cm^2 sec^{-1} at 300°C. From the above figures, it may be estimated that the time for H_2 to diffuse 1 mm through stainless steel at 300°C is about 40 hours, which gives an indication of the time required to degas a piece of stainless steel (the gas content of which is often mainly H_2) of this thickness. Exact calculation is complicated—see, for example, Lewin, *Fundamentals of Vacuum Science and Technology* (McGraw Hill Book Co., 1965), pp. 23 *et seq.*

9

§ 9. *Photons*

In some situations, the energy of electromagnetic radiation (e.g. light or X-rays) behaves as if it consisted of discrete packets of radiation, whose energy E is given by

$$E = h\nu$$

where　ν = frequency
　　　　h = Planck's constant
　　　　　= $6 \cdot 6 \times 10^{-27}$ erg seconds

§ 10. *Electronic structure of atoms*

An atom consists of a nucleus surrounded by a cloud of electrons (see § 1). The nucleus is positively charged and the electrons negatively, the number of electrons being such that the sum of their charges (see *Units and Constants*, § 2 above) is equal and opposite to the charge on the nucleus, so that the atom as a whole is, in its normal state, electrically neutral.

The electrons are not all similarly situated with respect to the nucleus; they have different orbits, those which are nearer to the nucleus being more strongly attracted to it and so more difficult to remove.

The removal of an electron from an atom is called *ionization*.

§ 11. *Ionization potential and efficiency of ionization*

The minimum energy (in electron volts) required to remove the most loosely bound electron from a gas atom or molecule is called the *first ionization potential* of that atom or molecule. The removal may be effected by electron impact, radiation (by a photon having an energy at least equal to the ionization potential), by temperature, or by a variety of other processes, provided they have the necessary minimum energy.

In vacuum technology, ionization by electron bombardment is of particular importance, being the basis of the measurement of pressure by ionization gauges.

The probability that electron impact will result in ionization depends on the energy of the bombarding electron. If the energy of the impinging electron is less than the first potential, the probability is zero; it increases with energy and reaches a maximum for most gases in the region of 100 to 200 electron volts, thereafter decreasing slightly with voltage. Hence the grid of a hot-cathode ionization gauge is usually between 100 and 200 volts positive with respect to the filament.

§ 12. *Thermionic emission*

The evaporation of electrons from hot solid bodies is known as thermionic emission. Like vaporization and outgassing it is a strongly increasing function of temperature.

It is of importance as providing the current of bombarding electrons in a hot-cathode ionization gauge, or for example in an electron beam welder.

§ 13. *Photoelectric emission*

Photoelectric emission is the emission of electrons from solids, liquids, and gases under bombardment by photons, i.e. electromagnetic radiation. In gases, the minimum energy of a photon necessary to cause ionization of the gas is given by

$$h\nu = eV_i$$

where V_i is the ionization potential. The frequency of the minimum energy photon necessary to cause electron emission from a solid is given by

$$h\nu = eV_w$$

where V_w is the work function of the electrons in the metal, i.e. eV_w is the minimum energy which an electron requires in order to escape from the metal.

§ 14. *Field emission*

Sometimes called *cold emission*, field emission is the emission of electrons from a metal at room temperature under the influence of an electric field normal to the surface of order 10^7 volts cm^{-1}.

§ 15. *X-rays*

When electrons bombard a solid, a small fraction of the kinetic energy of the impinging electrons is converted into photons, which are emitted from the solid and are termed X-rays.

X-rays of relatively low frequency, corresponding to energies of (say) $E = h\nu = 100$ electron volts, are called *soft*. They cannot penetrate glass, or air at atmospheric pressure, but have sufficient energy to cause the emission of photoelectrons from metals on which they fall, since the work function V_ω of metal is only a few volts.

Hard X-rays are generated by the impingement of electrons of many kilovolts energy. They are penetrating, and physiological protection requires the use of shielding; e.g. electron guns for welding and metal melting utilize electrons of energy in the 50 to 100 keV range and shielding against the consequent X-rays is necessary.

§ 16. *The gyromagnetic or cyclotron frequency*

If a particle, charge e and mass m, enters a magnetic field H having a component of velocity $v_\perp$ perpendicular to H, it is acted upon by a force orthogonal both to $v_\perp$ and to H and is therefore constrained to move in a circular arc, radius R, round an axis parallel to H, with an angular velocity given by

$$\omega = \frac{e}{m} H \tag{6}$$

The corresponding frequency, $\omega/2\pi$ is called the gyromagnetic or cyclotron frequency, the radius R of the orbit being proportional to the value of $v_\perp$ which gave rise to it.

If on entering the field H the particle has, in addition to $v_\perp$ a component of velocity $v_{||}$, parallel to H, the path will be a helix of constant pitch. If however there is an axial electric field E, parallel to H, the pitch of the helix will increase as the particle moves down the electric field and gains energy from it.

§ 17. *Cold-cathode discharge*

This may best be described with reference to a specific example. If an increasing D.C. voltage is applied between two plane electrodes spaced (say) 10 cm apart, the pressure being about 10^{-1} torr of air, a luminous glow discharge starts at about 300 volts. Its appearance is not uniform throughout its length, being different at the cathode and anode ends.

If the pressure is gradually decreased, the current decreases and the voltage required to maintain the discharge rises. Finally the discharge ceases, the current falling to zero and the voltage rising to its open-circuit value.

The mechanism of the discharge is as follows. The discharge is initiated by a few chance electrons, produced, e.g. by cosmic radiation or by the natural radioactivity of the surroundings. Under the influence of the electric field between cathode and anode, the initial electrons multiply themselves by ionization by collision with molecules of the gas, and the positive ions so produced impinge on the cathode to produce further electrons, which maintain the discharge. The total current measured in the external circuit is the sum of the positive ion current reaching the cathode, and the electron current leaving it and is a function of the gas pressure.

The above process, modified by the presence of a longitudinal magnetic field, is the basis of the Penning cold-cathode ionization gauge and of the sputter ion pump (§§ 4.6.1 and 3.4.1).

§ 18. *Sputtering*

This is said to occur when, in a gas discharge, the impingement of positive gas ions on the cathode causes uncharged particles of the cathode material to be emitted.

The phenomenon is exploited in sputter ion pumps to produce from the cathode a continuous supply of sputtered Ti atoms. These settle on the walls of the pump and 'getter' the gas. Sputtering is used to deposit films of high-melting-point low-vapour-pressure materials which are difficult to evaporate, e.g. Ta.

§ 19. *The Boltzmann factor*

In an assembly of molecules in thermal equilibrium at temperature T, the expression for the fraction that possesses, by virtue of thermal energy, an energy q is dominated by the factor $\exp(-q/kT)$ where k is Boltzmann's constant. The quantity q/kT is a measure of the ratio of q to the *average* kinetic energy of a molecule of the assembly. If $q \gg kT$, it will (since it occurs in the exponent of the exponential) cause the Boltzmann factor to increase rapidly with T.

Equation (1) of Appendix I exemplifies this. There $mc^2/2kT$ is a measure of the ratio of the kinetic energy of those molecules having a speed c to the *average* translational kinetic energy of the whole assembly, and the temperature dependence of the speed (and therefore kinetic energy distribution) is dominated by the term $\exp(-mc^2/2kT)$.

APPENDIX III
Additional References

Listed below are some useful references obtained too late to be included in the main text.

Chapter 2

LAWSON, J. D., 1966. Some calculations on the pumping of trapped volumes. *J. Sci. Inst.*, **43**, pp. 565–568.

Chapter 3

BAKER, M. A., and LAURENSON, L., 1969. The application of fore-line sorption traps. *Vacuum*, **19**, p. 81.

BROMBERG, J. P., 1969. Accurate pressure measurements in the region 0·1–1000 m torr; the inter-calibration of two McLeod gauges and a capacitance manometer. *Jour. Vac. Sci. and Tech.*, **6**, pp. 801–808.

HALL, L. D., 1969. Multi-component getter films for getter ion pumps. *Jour. Vac. Sci. and Tech.*, **6**, pp. 44–47.

HOBSON, J. P., 1969. Surface smoothness in thermal transpiration at very low pressures. *Jour. Vac. Sci. and Tech.*, **6**, pp. 257–259.

Chapter 4

MCCRACKEN, G. M., 1969. Small magnetic mass spectrometers as residual gas analysers. *Vacuum*, **19**, pp. 311–315. (The advantages and limitations of using small magnetic mass spectrometers as residual gas analysers are reviewed with reference to some commercially available instruments.)

Chapter 5

KREUCHEN, K. H., 1969. New ways with ceramics in electronic valves. *Proc. Brit. Ceram. Soc.*, March 1968, pp. 79–86. (The behaviour of ceramic surfaces is discussed with reference to gas occlusion, surface adsorption and desorption, and the influence of these on the working of the valve.)

MAIR, W. N., and BRAMHAM, R. W., 1968. A cheap sensitive leak detector. *Royal Aircraft Establishment Technical Report* 68099.

Chapter 6

PINGEL, H., 1968. Ultra-high vacuum system for a 3-GeV electron-positron storage ring at DESY, Hamburg. *Proc. 4th Internat. Vac. Congr.*, 1968. Institute of Physics and Physical Society. Conference Series No. 5, pp. 251–254.

Chapter 7

FEW, R. A., and CHERRY, S. J., 1969. Vacuum circuit-breakers for distribution systems. *Electrical Times*, 14 Aug. 1969, pp. 49–50.

GRUNERT, W. E., and NOTARO, F., 1968. Vacuum panel insulation systems. *Advances in Cryogenic Engineering*, **13**, pp. 690–700 (Plenum Press, N.Y.)

LEMPERT, J., 1966. The non-vacuum electron-beam welder as a welding tool. Scientific Paper 66-IC2-EWELD-P2. Proprietary Class III. Nov. 1, 1966. Westinghouse Research Laboratories, Pittsburgh, Pa. (Describes a welder in which the electron beam emerges from the electron gun into the atmosphere via differentially pumped orifices, the weld itself being made in helium at atmospheric pressure.)

HACMAN, D., 1968. Coating techniques in the U.H.V. range. *Bolzer Technical Information Sheet* 16, pp. 191–196. (Discusses problems of particular interest for the *industrial* applications of U.H.V. coating techniques.)

Bibliographical Note

Two important English language journals dealing exclusively with vacuum science and technology are:—

(i) *The Journal of Vacuum Science and Technology*, published bi-monthly for the American Vacuum Society by the American Institute of Physics, 335 East 45 Street. N.Y. 10017. U.S.A.

(ii) *Vacuum*, published by Pergamon Press Ltd, Pioneer House, 348 Grays Inn Road, London WC1, England.

APPENDIX IV

Some Chemical Elements used in Vacuum Technology

Name	Atomic weight	Chemical symbol	Temperature (°K) at which vapour pressure is 10^{-3} torr	Melting temperature (°K)	Boiling temperature (°K)
Hydrogen	1·008	H	6·0	13·9	20·3
Helium	4·003	He	10^{-1} torr at 0·98°K		4·22
Carbon	12·011	C	2520		4170
Nitrogen	14·007	N	34	63	77
Oxygen	16·00	O	40	54	90
Neon	20·18	Ne	11	24·5	27
Aluminium	26·98	Al	1355	933	2720
Argon	39·95	Ar	39	83·7	87·2
Titanium	47·90	Ti	1831	1953	3600
Chromium	51·97	Cr	1540	2176	2900
Iron	55·85	Fe	1595	1812	3100
Nickel	58·71	Ni	1630	1726	3100
Copper	63·54	Cu	1415	1356	2850
Zinc	65·37	Zn	563	693	1180
Molybdenum	95·94	Mo	2650	2893	4850
Silver	107·87	Ag	1195	1234	2450
Cadmium	112·40	Cd	494	594	1040
Indium	114·82	In	1100	430	2300
Tin	118·69	Sn	1365	505	2900
Barium	137·34	Ba	810	983	1900
Tantalum	180·95	Ta	3080	3270	5700
Tungsten	183·85	W	3280	3650	5750
Platinum	195·09	Pt	2180	2040	4100
Rhenium	186·2	Re	3060	3450	5900
Iridium	192·2	Ir	2380	2720	4700
Gold	196·97	Au	1470	1336	2950
Mercury	200·59	Hg	289	234	630

Notes

(1) The atomic weights are based on the 1963 values of the International Atomic Weight Commission, but are mostly rounded off to two decimal places.

(2) Temperatures are quoted in degrees Kelvin. To convert to degrees Centigrade add 273.

(3) The quoted boiling temperatures are, for most high-melting-point elements, obtained by extrapolating lower temperature experimental vapour pressures to the temperature at which they attain the value of 760 torr. They are, therefore, subject to uncertainty and merely give an indication of increase of volatility with temperature.

(4) Vapour pressure data for the more common elements and some common gases over a wide range of temperature are given by R. E. Honig, *R.C.A. Review*, **18**, June 1957, pp. 195–204 and *ibid.*, **21**, September 1960, pp. 360–368.

(5) The evaporation rate (E) *in vacuo* is related to the vapour pressure P of an element of atomic weight M, if it evaporates as single atoms, by the relation

$$E = 5 \cdot 8 \times 10^{-2}\, SP_{torr} \left(\frac{M}{T}\right)^{1/2} \text{ gram cm}^{-2}\text{ sec}^{-1} \qquad (1)$$

where T is the temperature in degrees Kelvin and S is the probability that an atom of the element, impinging on its own solid at temperature T, will stick to it. For most metals, S is of order unity, but may be considerably less for elements such as carbon, whose vapour is partly polyatomic.

(6) A useful table of the physical properties of the elements is given by A. H. Turnbull, R. S. Barton and J. C. Rivière in their book *An Introduction to Vacuum Technique* (1962, George Newnes Ltd). They also give a similar table (which includes information on weldability) for stainless steels and other alloys used in vacuum work.

SUBJECT INDEX

AUTHOR INDEX